BECOMING AI-READY

Designing Future Nonprofit Teams

Joe Manok and David Woodruff

ISBN: 979-8-9961239-5-7

Library of Congress Control Number (LCCN): 2026912492

First Edition 2026

Published by **Clark University Press**

Joe: To my mother, Hodhod

David: To my granddaughters, Emily and Sophie

With Insights from . . .

Alex Amouyel, Newman's Own Foundation
Amir Pasic, Indiana University Lilly Family School of Philanthropy
Amy Gingle, Orr Group
Ashley Budd, Cornell University
Bashshar Haydar, American University of Beirut
Brent Grinna, EverTrue
Cara Giacomini, CASE
Chelsea Lamego, FundMiner
Christopher Cannon, Generosity Collaborative
David Fithian, Clark University
George Bacher, Clark University
Iyad Rahwan, Max Planck Institute for Human Development
Jon Thompson, Children's Hospital of Philadelphia
Kerry Evanter, Clark University
Kim Infanti, Syracuse University
Kristen Fox, Business-Higher Education Forum
Mallory Erickson, Practivated
Mark Koenig, Oregon State University Foundation
Matthew Lambert, William & Mary
Matthew Letendre, Dana-Farber Cancer Institute
Maysam Ali, Amazon Web Services
Nathan Chappell, Fundraising.ai
Patrick Frontiera, Amazon Web Services
Russ Altman, Stanford University
Steve Suda, Stanford University
Taylor Swaak, Chronicle of Higher Education
Vuslat Doğan, Vuslat Foundation

TESTIMONIALS

"*Becoming AI-Ready* offers a thoughtful perspective for philanthropic leaders thinking about how emerging technologies may support their organizations. Its focus on judgment, values, and purpose is a useful contribution to an important conversation."

—Olivia Leland, Founder, Co-Impact;
Founding Director, Giving Pledge

"AI is a tool, not a strategy. *Becoming AI-Ready* reminds us that technology must always serve our core mission and values. This book offers practical guidance for nonprofit leaders aiming to embrace innovation without losing sight of the relationships and purpose that drive real impact."

—Ron Shaich, Author of *Know What Matters*;
Founder, Panera Bread; Trustee, Clark University

"The next generation of advancement will depend on the ability to translate insight into action while preserving trust. In *Becoming AI-Ready*, Joe Manok and David Woodruff demonstrate how AI can enhance the strategic work of fundraising while keeping relationships and values at the center."

—Eric Grimson, Computer Scientist and Chancellor for
Academic Advancement, MIT

"Institutions that aspire to endure must continuously adapt while remaining faithful to their mission. *Becoming AI-Ready* offers an important perspective on how AI can help mission-driven organizations strengthen strategy, governance, and long-term impact without losing the human values that sustain them."

—Abdo George Kadifa, Technology Investor; Chairman, Board of Trustees, American University of Beirut

"Having a roadmap to navigate the challenges and opportunities of adopting AI is essential for today's university chief advancement officer. *Becoming AI-Ready* offers an important perspective for leaders of nonprofits in building strong teams to meet the future."

—Brian Lee, Former Vice President for Alumni Affairs and Development, Harvard University

"AI and connectivity are the great accelerators of our time, reshaping every industry and every community. The organizations that thrive will be those that integrate deep technological capability with a relentless sense of purpose and responsible leadership. *Becoming AI-Ready* provides nonprofit leaders with a critical roadmap to navigate this transformation and ensure no one is left behind in the digital age."

—Hans Vestberg, Former Chairman and CEO, Verizon

"Fundraising has always been deeply human work built on relationships, courage, and trust. *Becoming AI-Ready* shows how AI can support fundraisers in that work, freeing them to focus less on administrative tasks and more on the conversations and connections that truly inspire generosity."

—Mallory Erickson, Co-Founder,
Fundraising.ai, Founder, Practivated

TABLE OF CONTENTS

FOREWORD
BECOMING THE BEST VERSION OF OURSELVES

Several years ago, I posed a question to our advancement leadership that was simple on its face, but expansive in its implications: When we look back 5 or 10 years from now, will we believe that we built the best version of ourselves? At the time, I could not have known how deeply that question would shape our thinking and our work. I asked it because I sensed that higher education, and our place within it, was entering a moment of profound change.

The expectations of students, alumni, families, donors, faculty, staff, and partners were shifting in meaningful ways. The pace of technological change was accelerating. Financial, demographic, policy, and societal pressures were converging, requiring institutions across our sector to reflect more honestly on how they operate and, more importantly, on what they stand for.

Yet I asked the question for another reason as well: because I believe deeply in Clark University as a place defined by purpose, imagination, and possibility. Clark has long challenged convention, not as an exercise in disruption for its own sake, but because the world needs institutions willing to evolve with intention, to experiment thoughtfully, and to lead with courage and clarity.

That initial question soon led to others. What would it mean for Clark to build a fundraising and engagement enterprise that truly reflects our values and responds to the way the world works today? How might we rethink our relationships, our practices, and our workflows so they prepare us for the future while remaining grounded in authentic human connection? And what would it look like to design an advancement model that is not simply better at the margins, but more fully aligned with who we aspire to be as an institution?

As we explored these questions, artificial intelligence emerged as a compelling place to begin. Not because technology itself was the destination (it never was), but because it offered an opportunity to think differently, to ask better questions, and to accelerate learning. AI was not our goal or our guiding star. It was a catalyst. From that starting point, the EverTrue-Clark Advancement AI Lab took shape as a collaboration grounded not in tools but in shared purpose and a collective commitment to becoming the best version of Clark Advancement.

This book tells the story of that journey, and it does so in the same spirit in which the work began. It is not a technical guide or a manual on emerging technologies. Instead, it is an invitation to reflect, to imagine, and to design. It invites readers to consider what the best version of their advancement office, their teams, and their organizations might look like: future-ready and mission-driven, yes, but also deeply human-centered.

At its core, a liberal arts education is not about telling people what to think; it is about teaching them *how* to think. We approached this work in that same way. Our hope is not to prescribe answers, but to help frame the questions, especially as we engage with technologies that are both promising and transformative.

As you read the pages that follow, I hope you will carry your own questions with you, reflect on what matters most in your work, and begin a journey of discovery that leads to answers shaped by your values, your mission, and your aspirations for the future.

—David Fithian
President of Clark University

INTRODUCTION
THE FRAGMENTATION PROBLEM

This book's publication is timely. Now at the tipping point of the Digital Age, there is ever-present technology that resides in our daily workflow and impacts our lives: artificial intelligence (AI). Although AI may be a polarizing topic, the technology is nonetheless becoming a reality for many organizations, including businesses, nonprofits, and universities. Two choices lie before us in these not-for-profit spaces: resist change and fall behind, or learn to ethically use AI to become more personalized. AI has no alma mater, no nostalgic attachment to a campus, and no mentor who changed its life. Not driven by emotion, AI cannot feel loyalty, inspiration, belonging, or mission. What it *can* do is support fundraising professionals by ingesting vast amounts of information, identifying patterns, and generating language that is personalized for each donor.

AI possesses both strengths and blind spots. It brings extraordinary pattern recognition and an objectivity unburdened by habits or fear of change. It neither clings to the way things have always been done in an organization nor favors one vendor, model, or workflow over another. At the same time, it cannot sense the mood of a boardroom, build trust over a cup of coffee, or empathize with a donor's life story (at least, not yet). Nonetheless, it is a powerful, fast, and increasingly useful tool, one that is here to stay.

As such, the institutions that will thrive in the years ahead are those that will learn to combine relational intelligence with artificial intelligence, using the latter to discover areas of improvement and opportunity, not to supplant judgment. Human values and intuition remain irreplaceable. During the COVID-19 pandemic, with the technology to reach anyone anywhere with video calls, we managed to connect across geographies and time zones in seconds, and yet once restrictions were lifted, we rushed to parks and backyards to connect with other people in person. This book doesn't suggest you replace your teams with AI; instead, learn where AI can support your team, freeing up their time to focus on the core work: the human part of philanthropy.

With that frame in place, we turn to the work that brought us here.

WHY OUR BOOK?

This point in history is a pivotal moment for mission-driven organizations. Leaders face diverging challenges with waning donor participation, even as institutional needs keep growing. At the same time, those who *do* give are expected to give more while demanding greater transparency and quantifiable impact. Today, public trust in institutions is fragile and must be continually earned while, concurrently, technological change accelerates faster than ever. This landscape of shifting participation, rising expectations, and fragile trust frames the central question behind our work: How can mission-driven teams strengthen their purpose and impact in an age of tools like AI?

The rise of generative AI in 2023 was met with unprecedented usage. Now, years later, every team needs to understand how AI will show up in daily work, yet many conversations in our field haven't kept pace with technology's leaps forward. At a time when trust is at a crossroads, how we integrate these tools into our organizations will have a lasting impact. That's where *Becoming AI-Ready* comes in.

Our first AI conversation took place in 2016 over lunch, during which we discussed Gravyty, a new startup focused on AI for fundraising and co-founded by the former frontline fundraiser and now serial entrepreneur, Adam Martel. We discussed the potential of machine learning and how it could enhance donor engagement. Who would have thought that exactly 10 years later, we'd publish a book together?

In the years since that first conversation, we have approached AI with a shared conviction that the real challenge is one of readiness. We met at MIT, where we worked together on a historic campaign, helping raise $6.2 billion for the university. The experience showed us firsthand the rising power of AI. One of us (Joe) moved from computer science into frontline fundraising at the American University of Beirut, then to MIT, and later to Clark University, co-founding the EverTrue-Clark Advancement AI Lab. The other (David) held advancement leadership roles at three research-intensive institutions: MIT, Harvard, and Massachusetts General Hospital, and today works as a consulting partner at Marts & Lundy and a trained leadership coach.

In these positions, we learned that being AI-ready means being organizationally ready, which requires cultivating the culture, talent, and ethical framework needed to harness new tools in the service of your purpose. Every day, we see leaders wrestling with a core tension: On one hand, there is pressure to move fast and innovate at the speed of technology. On the other hand, truly lasting progress requires trust, deliberation, and bringing people along at a different pace.

While AI promises efficiency and scale, we know that meaningful work still thrives on human relationships and empathy. This book addresses the heart of these tensions, outlining a balanced path of implementation over easy extremes. Our approach advocates for thoughtful adoption of AI, using it to augment rather than replace human reasoning and to enrich instead of erode the human-to-human connection that is at the heart of philanthropy. Too often, conversations about new technology swing between uncritical hype and paralyzing fear. We have watched nonprofits chase the latest tools, searching for a quick win or fix. We have also seen teams hesitate to change due to valid concerns around privacy, values, and other important factors.

Our focus is on sustaining strategic clarity and ethical rigor, illustrating how organizations can explore emerging technologies while prioritizing their purpose and stakeholders. Advancement offices can benefit from AI, but only if they invest in their people and culture, preparing them for the future. The foundation must be practical and human-centered: The future belongs to teams who are curious about innovation yet conscientious about its implementation.

WHAT THE READER WILL LEARN

In the following chapters, we examine what it means to be "AI-ready" beyond buzzwords. How does AI affect our sector? How does it impact culture, governance, data, ethics, and talent development? Alongside case studies showing how other organizations utilize AI, we also share frameworks that can help you evaluate which tools to use and what questions to ask, strategies to upskill yourself and your team, and guiding principles that worked for us, plus practical insights, so you can chart your own path to what works best for your organization. We present *Becoming AI-Ready* as a resource for visionary leaders, including advancement professionals, board members, emerging talent, and others interested in the role of technology and innovation in the nonprofit sector.

How We Used AI in This Book

Ethical AI requires transparency, so to practice what we preach, below is how we used AI for this book:

- We pressure-tested some ideas, including our outline and its market competitiveness.
- We used AI as a research assistant instead of using Google Search, including the ChatGPT Deep Research functionality.
- AI collected the URLs for our endnotes, fact-checked dates, and ensured consistency in the citation format. (We also fact-checked everything ourselves.)
- As we drafted the manuscript, the Microsoft Copilot feature was active, providing word and paraphrasing suggestions, with the spelling and grammar feature activated as well.
- We also used AI to produce a critique of our manuscript and provide light language refinement support.

All of this said, the proposed edits were reviewed and adjusted by us. The core of the book maintains the human touch that every fundraiser needs.

The path forward is bright. Undoubtedly, there will be more advanced tools in the years ahead and greater data-driven decision-making capacities, too. We are equally confident that nonprofits can and will rise to the challenge as they did with previous technologies over the past century. The next era of philanthropy will be defined by ethical guidance, responsible deployments, thoughtful leadership, a learning mindset, and intentional AI use. Assuming those elements all work hand-in-hand, not one without the other, we can build the best version of our industry: technically savvy, institutionally ready, and unquestionably human.

CHAPTER ONE

The AI Imperative

We find ourselves in especially challenging times for fundraising: Simply put, fundraising goals continue to rise, donor engagement is soft, advancement budgets are flat, talent wars and hiring freezes are at a peak, and finance teams warn of mounting cost pressures. This is the current climate, and AI is thought to be the answer. According to a recent Gallup-Lumina poll, 36 percent of Americans have confidence in higher education (down from 57 percent in 2015).[1] Public trust, the moral capital that universities and nonprofits depend on, is eroding.[2] This is exacerbated by the fact that unemployed recent college grads account for 12 percent of the rise in the national unemployment rate, exposing a stunning mismatch between educational promises and economic reality.[3] Students are better prepared than before, but many traditional early-career roles are disappearing. AI is thought to be the cause, but the reality is an overall change in the workforce. Association, not causation.

Many leaders hesitate to embrace AI for fear of increased audience alienation, but in doing so, they are all but guaranteed to fall further behind. The mounting obstacles require a drastic change: the AI imperative.

This is a moment of reckoning for mission-driven institutions, not only universities, but also nonprofits, charities, and foundations alike. Incremental tweaks won't regain public confidence or meet the rising tide of community needs.

What role can AI play amid all this change?

DECLINING PARTICIPATION, RISING GOALS, AND SHIFTING PUBLIC TRUST

Zoom out, and you see a wider convergence of trends challenging mission-driven organizations across the board. Traditional engagement models are faltering, enrollment in many colleges has been stagnating or dropping, and nonprofit membership and volunteerism are down. Philanthropic participation has been on a long-term decline. (The percentage of American households giving to charity fell from 66 percent to under 50 percent between 2000 and 2018.)[4] At the same time, institutions are expected to deliver more services, more personalization, and more innovation. This "do more with less" paradox is now the norm.

All of this would be challenge enough, but the external environment is changing in ways that amplify the strain. A report by the Business-Higher Education Forum (BHEF) highlights several powerful shifts, the most crucial of which is the reality that we are living in a technology-driven, AI age.[5]

One shift, in particular, is that the nature of work is transforming, with LinkedIn projecting that roughly 70 percent of core job skills are forecasted to change by 2030, largely due to AI and automation.[6] Training cannot be treated as a one-time event anymore. The preparation that universities provide and the on-the-job learning pathways that nonprofits rely on will both need constant updates. Routine, codified tasks are declining while work requiring judgment, interpretation, and complex human interaction is increasing, even in early-career roles. In effect, entry-level jobs are now demanding more skills from day one because AI is absorbing much of the routine aspect.[7]

Early-career pathways are fracturing, with young graduates facing tough markets and unemployment significantly above the national average.[8] Employers have pulled back hiring in "hot" sectors such as tech, while simultaneously using AI to streamline roles that used to be stepping stones.[9]

Employer expectations are outpacing educational models, as more than 80 percent of business leaders believe advances in AI will require entirely new skills and ways of working. At the same time, a large majority of organizations admit they are unsure how to proceed, with almost 9 in 10 saying they need external guidance on their AI strategy.[10] This paradox of high expectations for AI-fluent talent, along with low clarity on strategy, signals a growing capability gap.

The biggest risk for organizations may be not adopting AI fast enough, rather than the other way around. Tools that were once the secret sauce of Fortune 500 companies are now cheap and available on the cloud for free.

THE CASE FOR AI

If the first part of this chapter outlines the problem, this section offers elements of the solution. Instead of seeing artificial intelligence as a threat or a diversion, why should mission-driven teams embrace it now? In short, proper AI adoption can reduce widely felt stress and greatly enhance human talent.

When applied carefully, AI has the potential to improve learning, productivity, and well-being, according to a growing corpus of evidence from university studies to business tests. Many of these advantages are particularly important for charitable organizations and educational institutions with little funding.

Consider an experiment at the multinational corporation Procter & Gamble, which is well known for its consumer goods. Published in 2025, the study's outcomes were striking: One individual equipped with a generative AI (ChatGPT-4 at the time) could match the output quality of two full-time staff members not using AI tools for the modest cost of a software license.[11]

Equally important, AI-augmented teams performed best of all, suggesting that the ideal isn't AI replacing collaboration, but AI enabling individuals to contribute at a higher level and allowing teams to reach new heights. This could, of course, change as technology evolves, but as of now, this is the foreseeable future.

The most notable finding was that, in comparison with peers who did not use AI, employees who used AI reported noticeably fewer negative feelings and more positive ones, such as validation. In particular, younger employees expressed relief that AI could take care of time-consuming administrative duties, freeing them up to concentrate on growth and meaningful work.

Just a few years ago, this kind of AI support was unheard of; instead, the work was left to advanced machine learning experts to do at a very high salary, using costly hardware and software setups. Today, the days of imagining that gift officers could offload hours of data entry or initial prospect research to an AI helper are over. This is happening right now.

AI may also help to address one of the toughest problems we have noted: the loss of entry-level learning opportunities. Paradoxically, AI is starting to dramatically shorten the learning curve for complex skills while eliminating some junior tasks for those who do enter a field.

On the strategy front, AI can surface information, insights, and unconventional strategies that might augment your human expert strategies or change them altogether. This won't make everyone an expert overnight or make your junior staff act like veterans, but it will shorten the learning time and allow your AI-empowered team to add more value to you and your stakeholders. For mission-driven organizations worried about the talent pipeline, this could be a lifeline. New hires could become more competent quickly, and the absence of certain entry-level positions could be offset by AI-enabled onboarding.

A recent report highlighted that AI skills are commanding an average wage premium of 28 percent.[12] Investing in AI training would increase the competitive edge of your team and potentially double productivity for 28 percent of the cost. An Ernest & Young study shows that employers are missing up to 40 percent of AI productivity today due to the current AI talent gaps.[13] Consequently, many employers are starting to consider fundamental AI training as a foundational skill for future hires.[14]

The case for AI is about two strategies: an offense and a defense. On the offense, AI increases output, personalizes engagement, and amplifies your impact. On the defense, responsibly adopting AI is necessary for your team not to feel left behind. A critical step in identifying how to move forward is for your leadership team to have a clear AI readiness assessment.

The AI Readiness State (AIRS) Framework

Before rushing to implement chatbots or predictive models, leaders should pause and ask: How ready is my organization to absorb and leverage AI? Are people fearful or enthusiastic about it? Do we have pockets of expertise?

Are there executive support and governance in place, or would any AI project be a rogue skunkworks-like effort?

Developed by Clark University, the AI Readiness State (AIRS) framework is a simple yet powerful diagnostic tool that aligns with two other frameworks, one related to ethics and one for plotting AI readiness and ethics readiness. AIRS describes four levels of institutional AI readiness, from the most hesitant to the fully integrated. Think of it as a maturity model for your team that captures both technical adoption and psychological culture around AI.

These four levels are:

1. **AIRS 1: Reactive**
   The organization is skeptical, fearful, or unaware of AI. Little to no use of AI is happening internally, and any suggestion of it may be met with "not for us," or "not now." The prevailing mindset is one of avoidance or denial, and perhaps emerges out of concern that AI is just a fad or out of anxiety that it threatens jobs or values. Culturally, people in this category frame AI as antithetical to their mission or their advancement work. Leaders at this level often have not made any public statements or plans about AI.
2. **AIRS 2: Exploring**
   The organization has pockets of curiosity. A few staff members may be using a tool like ChatGPT on their own, or one department may have run a pilot in a controlled environment or for a one-use case. Leadership is aware of AI as an emerging issue but has not formalized anything. The mindset is a cautious experimentation with a mix of intrigue and concern. People are talking about AI in meetings or

forming small working groups, but there is no unified strategy or policy yet.

3. **AIRS 3: Operational**
   The institution has recognized AI as a strategic priority and has taken initial coordinated steps. There may be a draft of an AI policy or an official task force or committee. One or two formal pilot projects might be underway, with leadership engaged and parameters being set. This is what happens when an organization undertakes AI adoption. An example would be strengthening core operations, such as standardizing advancement workflows in CRMs like Salesforce to ensure data quality and readiness for responsible AI adoption.
4. **AIRS 4: Strategic**
   At this stage, institutions treat AI as a core driver of mission, fully aligning it with planning, budgeting, and leadership priorities. AI is no longer experimental; it is integrated into daily operations and guided by formal policies that cover ethics, data use, procurement, and oversight. In this category, the organization focuses on the "i" in AI, augmenting human intelligence with machine intelligence and making the institution smarter, more agile, and strategic. Multiple departments use AI confidently with documented wins, shared learning, and AI training embedded into onboarding and operations. This is when AI becomes part of the culture and strategy rather than its own initiative or task force.

Why does determining your AIRS level matter? Former CEO of Panera Bread and author of *Know What Matters*, Ron Shaich, knows that strategy starts with telling the truth.[15] AIRS 1 should not attempt to implement an advanced AI-driven personalization without getting into AIRS 2. We will revisit AIRS, particularly in chapter nine, but for now, remember its role as a diagnostic tool and the mirror at the beginning of the transformation.

Expert Voice: Amir Pasic, Ethical Leadership in an Algorithmic Age[16]

Amid the talk of technology and strategy, it is essential to ground thinking in ethics and humanity as the true north for mission-driven work. For that, consider the perspective of Amir Pasic, an advancement leader and the dean of the Lilly Family School of Philanthropy at Indiana University. In this narrative, Pasic serves as an ethical compass, a reminder that *how* we innovate is as important as the choice to innovate.

Pasic offers the cautionary reminder that philanthropy's work is not merely technical; it is also moral. In his reflections on AI's rise, Pasic stresses that intelligent systems are powerful tools, but their direction must be governed by the same human values that have long been anchored in generosity. AI is a transformative technology that demands its own ethical framework, rooted in humane imagination that philanthropy specializes in cultivating. Advanced tools may draft messages or analyze vast datasets, but only people can ensure that these systems reflect the values of compassion, responsibility, and public purpose.

Pasic said that fundraising, as a profession, extends beyond solicitation and into the freedom to question and connect with others about causes they find important. Unbeholden to donor interests, fundraising remains unwavering in its commitment to mission and values, and that independence, he notes, is essential to maintaining trust and protecting the integrity of mission-driven work. Trust is the throughline in Pasic's work, and although public confidence in institutions has eroded, he points out that nonprofits still enjoy comparatively high, although still fragile, trust compared to other industries. Simply put, tools like AI should amplify human judgment, not override it.

NEW LEADERSHIP TENSIONS

AI adoption will revolve around balancing three key tensions.

1. **Speed vs. Trust**
   Unlike the for-profit world, where moving fast is often rewarded, the nonprofit world intentionally moves more slowly and has to regularly balance speed with trust. It took our predecessors decades to build trust with the communities that we serve, and there is no amount of AI efficiency worth risking that trust.
2. **Efficiency vs. Humanity**
   AI is often sold as a tool for efficiency that reduces costs, eliminates redundancy, and optimizes resources. For cash-strapped organizations and tuition-dependent colleges, this is tempting. Many donors support these nonprofits for the human touch: the "thank you" call from a scholarship

recipient, the handwritten note for contributions to a fundraiser, or the coffee with the president. AI, if implemented carelessly, can strip away these moments in the name of optimization. One approach is to let AI handle the background work and allow the frontline to handle genuine human interaction.

3. **Innovation vs. Integrity**
   AI opens doors to new ways of solving problems and creating value. Innovation means continual experimentation, agility, and willingness to try bold ideas. AI drives a pace of change and a temptation to push boundaries that can clash against institutional values. Leaders will constantly face "can we" versus "should we" decisions. One practical technique is to establish "red lines" of actions we should never take, even if technically possible, and "green lines" of areas where use is encouraged. Later in the book, we discuss the value of writing your personal position statement on the use of AI, which is essentially a personalized list of "red lines" and "green lines" that you are comfortable with.

Across all three tensions, the goal is not to choose one side or the other, but to continually balance and rebalance. The focus should be on creating a culture where trade-offs are open for discussion. Those who will hold these dualities of being careful, efficient, empathetic, and innovative yet principled will steer their institutions successfully through the AI era.

WHY IS THIS MOMENT DIFFERENT?

Organizations have weathered waves of technological change before, from personal computers to the web to mobile. What makes this moment in time distinct are the following factors:

- **The cost of AI adoption has plummeted, and accessibility has skyrocketed.**
  AI was previously restricted to PhD experts in machine learning and so forth, but now, any small nonprofit or academic department with an internet connection can leverage AI, thanks to cloud computing and cheap, publicly available models. GPT-5 models (at the time of writing this book) are accessible for pennies per query, and open-source models can run on a standard laptop or your phone.
- **The cost of inaction has risen.**
  With previous tech shifts, an institution could sometimes wait on the sidelines and join later without catastrophic consequences. AI feels different due to its fast uptake and compounding advantages; sitting out even a couple of years could widen a gap that is hard to close. Organizations that started between 2018 and 2022 are far ahead of those starting in 2025. Public perception is also shifting, with most donors expecting personalization and treatment equal to what they are receiving from other nonprofit and for-profit organizations they work with.
- **Talent expectations are permanently changing.**
  The incoming generation of workers and the growing share of current ones are already starting to use AI in their personal and academic lives. They expect

their workplace to do the same. Organizations with forward-thinking AI practices are likely to attract talent that is innovative, goal-oriented, and mission-driven. In turn, AI readiness is becoming a key part of organizational attractiveness and culture.

On top of these factors, AI has become pervasive in daily life. From website chatbots for customer experience to recommendation engines, people are getting used to a world where AI plays an assistive role. Just as having a functional website became nonnegotiable 20 years ago, having AI-enhanced responsiveness and personalization will soon become standard.

This moment feels like an inflection point, and those who lean in will set the norms and models for the sector. Those who lean out may find themselves scrambling to catch up or wondering why support and engagement continue to dry up.

If you have not started your AI journey yet, you have plenty of time to do so, but the sooner you start, the sooner you can make that critical progress.

For us, crossing the threshold means having a clear plan, starting with a limited pilot or having a candid conversation with your team. It is not about rushing to the use of AI, but initiating a thoughtful process by which you start thinking about what it means for you to use it. Our hope is that over time, you will look back at this threshold and see your progress that will help you transform and not just "survive" the AI wave.

Chapter One Recap

Core Message

Mission-driven institutions are experiencing a convergence of pressures from declining trust, rising expectations, shrinking talent pipelines, and accelerating technological change. AI is no longer optional. Instead, it's a strategic lever that enables nonprofits and universities to remain effective, relevant, and trustworthy. Responsible adoption requires both leadership and cultural mandates. The path forward begins with honesty about readiness, clarity of purpose, and a commitment to ethics and trust as the foundation for all innovations. AI is not an experiment, a hype, or a technical project. It is a strategic, ethical, and cultural transformation.

Key Takeaways

- Public trust in higher education and nonprofits is eroding, donor participation is declining, early career job markets are unstable, and expectations for personalization are soaring. Institutions must adapt or risk further disconnection.
- Workforce shock is redefining talent.
- The case for AI is dual: offense and defense. AI helps teams produce more with less, and not adopting it poses long-term risks in talent, relevance, and mission delivery.
- Ethics and values remain important to protect public trust.
- Leadership tensions define the AI era, and we must be prepared to continuously balance them.
- AI is cheaper and more accessible than ever.

Chapter Endnotes

1. Jocelyn Gecker, "Is College Worth It? Poll Finds Only 36% of Americans Have Confidence in Higher Education," *AP News*, updated July 8, 2024, https://apnews.com/article/higher-education-confidence-survey-574e02d8478d3c1a7567a54f61af11f1.
2. Jeffrey Jones, "U.S. Confidence in Higher Education Now Closely Divided," Gallup, July 8, 2024, https://news.gallup.com/poll/646880/confidence-higher-education-closely-divided.aspx.
3. Megan Cerullo, "Rising Number of College Grads Are Unemployed, New Research Shows," *CBS News*, updated May 28, 2025, https://www.cbsnews.com/news/college-graduate-unemployed-technology-artificial-intelligence/.
4. Rebecca Klapper, "Number of Americans Giving to Charities Declined to 49.6 Percent, Study Shows," *Newsweek*, July 27, 2021, https://www.newsweek.com/number-americans-giving-charities-declines-42-percent-study-shows-1613539.
5. Candace Williams and Kristen Fox, "What's Next for AI and Digital Skills in the Workforce," Business-Higher Education Forum, March 12, 2024, https://www.bhef.com/article/2024/whats-next-for-ai-and-digital-skills-in-the-workforce.
6. *Work Change Report: AI Is Coming to Work* (LinkedIn, 2025), https://economicgraph.linkedin.com/content/dam/me/economicgraph/en-us/PDF/Work-Change-Report.pdf.

7. *The Future of Jobs Report 2025* (World Economic Forum, 2025), https://reports.weforum.org/docs/WEF_Future_of_Jobs_Report_2025.pdf.
8. Taylor Telford, "Facing Entry-Level Job Crunch, New Grads Question the Value of a Degree," *Washington Post*, June 26, 2025, https://www.washingtonpost.com/business/2025/06/26/graduates-degree-value-job-market-ai/.
9. *Work Change Report: AI Is Coming to Work* (LinkedIn, 2025), https://economicgraph.linkedin.com/content/dam/me/economicgraph/en-us/PDF/Work-Change-Report.pdf.
10. "The State of AI in 2025: Agents, Innovation, and Transformation," McKinsey & Company, November 5, 2025, https://www.mckinsey.com/capabilities/quantumblack/our-insights/the-state-of-ai.
11. Charlie Cowan, "P&G, Harvard Business School AI Study," CharlieCowan.ai, March 25, 2025, https://charliecowan.ai/blog/pampgharvard-business-school-ai-study-new-research.
12. "New Lightcast Report: AI Skills Command 28% Salary Premium as Demand Shifts Beyond Tech Sector," Lightcast, July 24, 2025, https://lightcast.io/resources/blog/beyond-the-buzz-press-release-2025-07-23.
13. Renny Popoola, "EY Survey Reveals Companies Are Missing Out on up to 40% of AI Productivity Gains Due to Gaps in Talent Strategy," *EY Newsroom*, November 10, 2025, https://www.ey.com/en_gl/newsroom/2025/11/ey-survey-reveals-companies-are-missing-out-on-up-to-40-percent-of-ai-productivity-gains-due-to-gaps-in-talent-strategy.

14. Andrea Willige, "AI Skills Are Becoming More Important than Job Experience. Here's How Job Seekers Can Adapt," World Economic Forum, June 6, 2024, https://www.weforum.org/stories/2024/06/ai-skills-rival-job-experience-in-recruitment/.
15. Ron Shaich, *Know What Matters: Lessons from a Lifetime of Transformations* (Harvard Business Review Press, 2023).
16. Amir Pasic (dean, Indiana University Lilly Family School of Philanthropy), interview by Joe Manok and David Woodruff, October 2025.

CHAPTER TWO

Fundraising's Four Data Frontiers

In the not-so-distant past, advancement teams depended on monthly static reports. Often presented in Excel, we curated reports, tracked the pipeline, and created year-end donor lists; these served as our compass to "do, track, and measure." These reports told us what happened, but by the time they were produced, in many instances, they were already out of date. Subsequent questions required days to get answers and additional time to rerun the reports. This retrospective approach can be compared to navigating your car using the rearview mirror.

In the AI era, the present and future of fundraising is about real-time insight informed by data and humans together, evolving strategy and outcomes instantaneously. Using the car metaphor again, imagine having 20 cameras surrounding you as you drive, evaluating your every move, and providing live insights when necessary, from a beeping sound when another car is closing in, to a view of a specific angle when you signal, and so forth. From the nonprofit perspective, identifying who is signaling interest, primed for engagement, open to being asked, or seeking their next opportunity transforms us from having to rely on reactive reports to being able to produce proactive, donor-driven strategies and responses.

Instead of waiting for an analyst to compile an end-of-month report, a relationship manager can receive a notification the moment an alumnus gets a big promotion or when a news article highlights one of their professional endeavors or family events. When data streams in from a wide range of sources in a timely and continuous manner, our outreach and interactions become more personalized, prompt, and relevant. We can stop asking what could be happening and start knowing what *is* happening. Rather than upgrade what we already produce, this is an opportunity to fundamentally change our mindsets and operations, to react in minutes instead of months.

For example, we used to meet once a quarter to review our donor insights. We would figure out who remained a prospect, who or what needed to be prioritized, if anyone made a gift, etc. Now, at almost every one-on-one, we reevaluate and prioritize specific people and projects. Because the insights are being received daily, or even hourly, we have to adjust our mindset accordingly to match the speed of incoming information.

Sales and marketing teams across the for-profit sector use this model today. According to BCG estimates, the market for companies that understand personalization will be in the range of $2 trillion over the next five years.[1] Teams get instant alerts when a prospect clicks on a link or when there's a change in a customer's behavior. They can act on these signals they receive almost instantly, with research from McKinsey showing that personalization could drive up to 15 percent in revenue.[2]

Instead of having the same language on, say, an alumni newsletter used for each recipient, you can enhance the interaction, speaking directly to "James" or "Sydney" with a detailed, personalized message. This extra step can make donors more likely to engage with the content sent to them.

Fundraising is starting to tap into this power. Imagine that you and your team are not looking at last week's static dashboards but rather into the dynamic feed of donor signals curated particularly for you. The technology then shows you which lapsed donor just wrote to the annual giving email or attended an alumni event, which major prospective donor opened your email twice, or which corporate stock surge might impact the behavior of some of your donors who work there.

Embracing these real-time insights means breaking the habit of "report and react" and reimagining our workflows accordingly. It means asking colleagues to proactively investigate these insights and not wait for reports to come to their inboxes to trigger a reaction. This change could allow us to have a live panoramic view of donor engagement.

The goal of this chapter is to provide you with a roadmap for making this mindset shift happen. We introduce a streamlined, effective model, created by the EverTrue-Clark Advancement AI Lab, to integrate all those sparks of data across different kinds of information that we see and harness the power of AI to transform them into a guiding light.[3] By the end, "real-time insight" will go from a buzzword to a practical reality. We will observe how connecting the dots into actionable intelligence, faster than ever before, helps us navigate the future of fundraising with confidence.

Expert Voice: Brent Grinna[4]

"The next era of fundraising will belong to organizations that break down silos and listen across all Four Data Frontiers."

—Brent Grinna, Founder and President of EverTrue

When I first started EverTrue, I was trying to solve a frustration I felt personally as a volunteer fundraiser and one I watched advancement professionals at my alma mater, Brown University, wrestle with constantly.

As a first-generation college graduate and financial aid recipient, I was deeply appreciative of the access to an education and network that I couldn't have imagined growing up. I wanted to give back. As vice president of the Brown Club of Chicago and a class agent, I saw how hard it was to do meaningful work with outdated spreadsheets and disconnected systems, and how difficult it was to match alumni passions with institutional needs. Advancement teams were doing important work, but they were doing it despite their tools, often navigating almost blind.

Fast forward more than a decade. Access to data has grown exponentially, yet many advancement leaders tell me they have less confidence than ever about what to do next. Donors increasingly expect the kind of relevance and personalization they experience from Amazon or streaming services, but what they often receive instead is generic outreach that feels disconnected from who they are and what they care about.

We have more signals than ever, but far too little understanding.

Our systems were never designed to talk to one another. Prospect researchers pore over wealth indicators. Annual giving teams track email open rates. Gift officers log notes in the CRM. Deans circulate priorities and wish lists by email. Engagement teams track event attendance. Athletics programs drive ticket sales. Each group holds a piece of the puzzle, but no one can see the full picture.

This is needle-in-a-haystack fundraising. One team hunts for a donor's passion in one haystack. Another hunts for a project in another. And we hope that by the end of the fiscal year, the needles somehow collide.

The truth is: The answers exist in the data. The signals are hiding in plain sight. The challenge is arranging them so they can speak, so they can tell us who is leaning in, what they care about, and how they want to engage, today.

I experienced this personally, and you can, too. I asked ChatGPT to review all publicly available funds at Brown University and recommend one that might resonate with me, using only publicly available information about me and about Brown's programs. It immediately gathered results of rural students at Brown, including initiatives like the STARS@Brown fly-in program, which helps high school students from rural communities overcome travel and access barriers to visiting campus.

I grew up on a 160-acre farm in Iowa, with all the formative experiences that entailed: raising pigs and cows in the 4-H club and learning parliamentary procedure in the Future Farmers of America (FFA).

The alignment was obvious and deeply meaningful. Yet in more than 20 years of engagement with Brown, including serving on the board of both the alumni association and the Brown Football Association, I had never been introduced to this fund. I wouldn't expect any fundraiser to know every fund or make every connection manually, but that experience crystallized something important for me.

The future of philanthropy cannot depend on donors having to find the right fund. Philanthropic opportunities that align with a donor's firsthand experiences, values, and identity should find the donor. And every fundraiser should have that kind of guidance at their fingertips.

This is why I believe the next era of fundraising will belong to organizations that can break down silos and listen across all Four Data Frontiers:

- private donor data,
- public donor data,
- private institutional data, and
- public institutional data.

The most effective advancement strategies and the best donor experiences will emerge at the intersection of these four frontiers. When organizations learn to listen across them, fundraising becomes both radically more efficient and more authentically human.

A COMPASS FOR THE AI ERA: FOUR DATA FRONTIERS EXPLAINED

In order to replace luck with insight, we created a model that examines the Four Data Frontiers. Think of this model as a compass for the AI era that aligns the data driving our decisions, mapping all advancement data along two axes:

Private vs. Public: Does the information live in the open or behind your institution's walls? Private data is information locked behind your firewall. It's the data in your donor information system, giving records, event attendance, wealth indicators, a summary of interactions with senior leaders and relationship managers, proposals, and internal documents that are not published.

Public data includes anything that is openly available for anyone to see: news articles, social media, press releases, real estate records, publicly available giving lists, board memberships, rich lists, and so forth. In a later chapter, we highlight that from a legal perspective, this information being in the public domain doesn't mean that an AI model can automatically review it. Particularly at the time of writing this book, the European Union and the State of California in the United States had restrictions on what could be collected and stored.

Donor vs. Institutional: Does the data describe an individual or organization (donor, prospective donor, alum) or your institution's priorities and context, such as offerings, new initiatives, faculty research projects, ranking, board reports, website, and so forth?

Placing the two axes perpendicular to each other gives us four quadrants: the Four Data Frontiers of modern fundraising. Every advancement office, whether they realize it or not, has information in each of these quadrants.

The Four Data Frontiers are plotted on a two-by-two matrix of private/public and donor/institutional data. By mapping our data this way, we gain a clear view of our blind spots and opportunities for information enhancement.

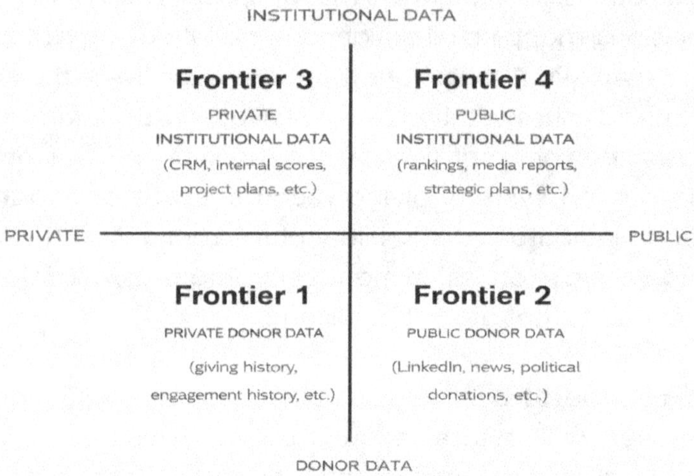

Four Data Frontiers developed by the EverTrue-Clark Advancement Lab

Many universities excel at collecting private donor data but do very little with publicly available donor data, like career moves, social media engagement, or giving to other institutions. Such data is considered human-heavy to collect and requires research professionals to identify, research, fact-check, and edit. These research professionals were reserved for presidential prospective donors or senior leadership, as they have limited time to produce such reports that require hours to generate. We're seeing that AI is not replacing these skills, but rather allowing us to provide this presidential-level research to all our relationship managers. Our research colleagues can relay this information to our relationship managers compiled in the format that they prefer in minutes. Then, they can have an AI system evaluate it against laws, policies, and codes of ethics, and allow our research colleagues to fact-check and edit them.

Other institutions might be good at leveraging public institutional data, but not so good with private institutional data, soon-to-be-announced initiatives, or yet-to-be-funded projects. The Four Data Frontiers force us to ask: Are we listening to all the signals, or just a select few? Each frontier by itself is extremely valuable and useful, but the key is integrating them together in a meaningful way using AI so that our best signals don't get lost or overlooked.

We have the full history from a donor to our institution (private donor data) in our database, but unless we connect the history with a trigger from the outside, saying their company is doing well, we might not recognize that we should accelerate an ask. Or we might have a list of funding opportunities (private institutional data) that were compiled at some point and are now gathering dust somewhere due to a lack of donor interest at the time. Now, due to national, international, or institution-specific events, these ideas are gathering alumni interest, but no one is matching those signals to the list. Amazing things can happen when we listen to all four quadrants. Individually, they provide important insights, but combined, they're catalytic. This is how transformational gifts, or the largest gifts of a lifetime, start to emerge by design and not by accident, as they sometimes do. In the next section, we will take a deeper dive into each of the Four Data Frontiers with examples of how nonprofits can move from report to insight to action by combining these streams of data.

Quadrant-by-Quadrant Examples

Let's explore each of the Four Data Frontiers one at a time to see what they entail and how connecting them can lead to some real wins.

Frontier 1: Private Donor Data

Key question: What has this person already done (or signaled) directly in their interactions with you?

The answer may help predict future behavior, provided that you review the information. This often sits internally waiting for a relationship manager (a frontline fundraiser) to discover it, mostly when they start managing the relationship and survey historical data.

Typical sources: Look at everything in your donor information system that relates to an individual. This is essentially the history of the donor engagement with your institution, which includes, but is not limited to, past givings, past pledges, event attendance, volunteer roles and board memberships, contact reports from gift officers of their key interactions, alumni network or career portal interactions, responses to surveys (if the responses are not anonymous), biographical information, and degree information.

Example: Nonprofit A decided to take an inward look at its data for insights, and it discovered overlooked information from years of volunteer forms and event follow-up notes. Using an AI model trained to detect language patterns of mission alignment, gratitude, interest in deeper involvement, or appreciation of their time with the organization, the team identified hundreds of individuals who had quietly signaled their readiness to commit at higher levels. By surfacing some of these hidden cues, Nonprofit A could increase the pool of potential major donors assigned or in its portfolio without adding names to its database. What had once been buried in handwritten notes and comment boxes became more of an actionable intelligence. Taking quick photos of these handwritten notes and analyzing the handwriting and content became something that can happen in seconds. In a pre-AI world, no one had the time to manually read through thousands of remarks. Today, the algorithms can highlight these areas where someone expressed interest in something or an attendee wanted to do more. This frontier turns institutional memory into a living archive of donor intent waiting to be rediscovered.

Private donor data does not have to come only from your donor information system. Most organizations use platforms like LexisNexis, which aggregate billions of records from corporate filings to philanthropic gifts and legal disclosures. These platforms are used by prospect researchers to confirm affiliation, identify opportunities or reputational risk, and help build a fuller picture of a donor's professional, civic, and wealth footprint.

At the next layer are predictive systems like BWF AI or DonorSearch AI (which is now part of the EverTrue ecosystem). These platforms use similar available filings and millions of records, but add a predictive analytics component to help analyze this data and help predict who is likely to engage, attend an event, support philanthropically, accept a meeting, or give at a higher level, all while analyzing private (but not personal) data. DonorSearch underscores its value in an article titled "AI for Nonprofits: How to Harness Its Full Potential," stating, "As of 2024, 58 percent of nonprofits have incorporated AI into their communications, and 68 percent use it for data analysis, both of which exceed for-profit implementation rates."[5]

The goal of these predictive analytics is not to replace human intuition but to amplify it and focus on what matters most: transforming years of static records into forward-looking insights that help relationship managers act sooner, with more empathy, personalization, and relevance.

Frontier 2: Public Donor Data

Key Question: What signals is this person sending out into the world that might matter to your organization? Are they signaling a passion, a life change, a new capacity, or a connection that you should note?

Typical sources: The donor will not share this information with the institution privately. Rather, they will put this on the internet or on other public forums: news articles, public social media profiles, press releases, corporate websites, board memberships, real estate records, publicly visible philanthropy, and political contributions.

Example: Imagine a mid-sized organization, Nonprofit B, that wanted to make donor engagement more timely and personal. Its fundraising team decided to incorporate publicly available professional announcements from social media and other public data sources into its fundraising workflow. Every time someone affiliated with the organization took on a new role or got a promotion, the system automatically flagged it. As you know, a career milestone is usually considered a moment of personal pride and most likely a capacity increase moment. This is a great opportunity for Nonprofit B's leadership to send a warm, personalized note to congratulate the donor.

Public donor data, when used thoughtfully, can transform casual digital signals into opportunities for authentic engagement. The key here is attentiveness and not surveillance or automation. Exhibit A in the appendix incorporates an illustrative "Super Prompt" that can be used in ChatGPT or similar products to generate an in-depth report of what is publicly available. This requires only sharing the first name, last name, and company of the prospective donor, and the output is often a 15- to 20-page report. At the time of writing this book, such reports would cost 90 cents on ChatGPT-5 and take around 12 to 15 minutes to produce using their deep research feature.

Disclaimer: These prompts must be reviewed by your general counsel and an oversight committee to make sure that any information you are collecting and/or storing aligns with the laws of where your organization is based, the laws where the donor is based, and your institutional values. A later chapter is dedicated to discussing ethics and regulations around data collection and privacy.

Frontier 3: Private Institutional Data

Key Question: What opportunities, needs, ideas, or projects are you trying to advance in that donors might align with?

In many cases, relationship managers are not fully aware of all the potential opportunities and exciting things happening at their institution. Information is changing rapidly, with data being fragmented across silos. Without centralization or a clear owner, valuable insights tend to be missed or unnoticed. Without surfacing these details, we won't be able to match the best win-win scenario for donor impact and institutional priorities.

Typical sources: This is basically the repository of information about what the institution might achieve if resources were allowed and available, such as internal planning documents, reports, and memos that are not yet in the public domain, as well as strategic plans, campaign priorities, case statements, meeting minutes, wish lists from faculty or from department heads or program and institute directors, unpublished proposals, and research initiatives that need funding.

Example: A compelling example of using private institutional data comes from University C. During a recent campaign, the university's advancement team conducted a structured mapping exercise of all faculty research priorities and projects emerging into one place. Much of this data already existed, some in departmental memos, others in internal planning documents, and some in proposals to potential federal funders and large foundations. They then cross-referenced these priorities with their donor information systems and identified, based on past affinities and interests, that some shared themes and potential prospective donors to these priorities. This approach surfaced more than 20 potential donors who were previously overlooked due to their narrow interest in very specific areas that the institution's faculty did not cover at the time, and were marked as "disqualified." Due to how the systems work, what the institution "needed" and what the prospective donors "publicly cared about" were not matched. Private institutional data, when thoughtfully analyzed, can serve as a significant matchmaking engine for philanthropic support.

Frontier 4: Public Institutional Data

Key question: What external perceptions or events, such as the institution's public narrative, successes, or struggles, are affecting the institution's appeal?

Typical sources: Imagine this information as a mirror held up to how others see the organization, how they talk about it, and how it presents itself to the world. This is outward-facing information about the institution or its environment: the university's website, news page, rankings and ratings (like the US News and World Report ranking page), media coverage, reputation scores or awards, public statements or speeches by leadership, buzz in the community, social media sentiments, government data, and public reports about the institution.

Example: Recently, a colleague who works at the foundation of a state university, University D, reported that when they had a big jump in their national ranking in the US News and World Report, which was preceded with two years of turbulent times, leadership transition, and challenging public image, they quickly launched a targeted outreach to boost morale with their lapsed donors from previous years who didn't give SYBUNTs ("Several Years But Unfortunately Not This Year"). They reported, after that story, a 24 percent increase in reengagement from those lapsed donors, indicating that nearly a quarter of these disconnected donors reconnected with them. Public institutional data can be a driving force or a hindering force for giving.

Bringing the Frontiers Together

The above examples highlight each quadrant in action, but the magic truly happens when you link all the quadrants together at once and look holistically at the donor experience, donor relationship, what they are signaling publicly, and what they are sharing with you privately, matched with what is said about your institution in public (and how it presents itself publicly) and your private reports, needs lists, strategic frameworks, and planning documents.

In our immediate manual past, we often had these data frontiers in monthly updated spreadsheets, reports sitting on the server begging to be read, or information on websites we rarely browsed or remembered. The AI-powered present is allowing us to connect these streams in real time and with higher precision so that we personalize and customize the donor relationship with us in a way that can provide the best possible experience to our donors and generate the best win-win outcome for both the donor and the institution. An easy way to remember this is:

Manual past = monthly spreadsheets and missed timing
AI present = precision outreach and real-time alerts

When all four frontiers are monitored, our haystack of data turns into magnets for actionable insights. We go from searching for needles to letting magnets (integrated data and AI insights) attract them and bring them to us.

The following section will show how to ignite those sparks systematically so that these "collisions" of data happen by design. In a way, the information that follows is a practical playbook for bringing the Four Data Frontiers to life for your organization.

When the Four Data Frontiers Align by Chance

If you've been around the fundraising profession, you can probably think of a handful of stories like the one below.

In the advancement office of University X, a gift officer, let us call her Rebecca, was cleaning her office and came across a folder with some donor information in it that she had previously printed and kept under lock. The folder included notes from an alumni panel that the donor participated in on campus, a scholarship donation from years before, and a handwritten note from an alumni relations colleague that said: "interested in student research."

During the same week, a colleague was working on giving priorities of an urban planning-related initiative with a wish list for funding an urban climate resilience lab that was shelved due to a lack of potential donors. Later that month, a news release about a faculty member who won a national award for urban heat mapping was published by the marketing and communications team, a relatively small news item for a large institution.

Six months later, on LinkedIn, the alumna posted a story about her company's work on equitable climate technology, and a colleague of Rebecca's shared that with her. At this point, you can most likely guess the rest of the story. Rebecca forwarded the information to her colleague at the dean's office, and the rest is history: a six-figure gift closed within eight months.

The same thing happens in every industry. An exceptional professor at MIT, Regina Barzilay, is known for her work on machine learning and old languages and manuscripts. She told her story publicly a few years ago about how she was diagnosed with cancer, and the highly reputable medical center couldn't answer specific questions she had about her treatment options.[6] It took an AI expert with personal experience with a specific disease to realize that the medical center, which was sitting on hundreds of thousands of data points and high-resolution images, was giving basic recommendations for the detection and treatment of something as life-threatening as cancer.

WHAT CAN AI TEACH US ABOUT MAKING LUCK REPEATABLE?

In Barzilay's case, she built an AI lab, the Abdullatif Jameel Clinic for Machine Learning in Health, that made the serendipity of her story much more thoughtful and systematic.[7] The Four Data Frontiers can help make that happen for the anonymized story above for Rebecca. Human intelligence takes time, effort, and, in the case above, a lot of luck. Luck will still play a role in our work, but we can curate a story like Rebecca's in a month using AI, rather than in 20-plus years. Colleagues considered Rebecca lucky as she had four bulbs lit at once. But it was a manual version of what modern systems can do automatically. The main issue is not luck but probability.

As Barzilay's work at MIT illustrates, the real breakthrough often comes from re-listening to ordinary data. Barzilay didn't come up with new medical imaging devices or new treatments, but rather transformed routine images, once considered too common, to reveal any new insights into predictive models that detect early cancer risk months or even years in advance. Her story reminds us that innovation is rarely about how to collect more data; it's sometimes about seeing the patterns hiding in plain sight.

In fundraising, that same mindset turns forgotten notes, overlooked fields, and dormant Excel files into forward-looking intelligence that can strengthen human decision-making.

The convergence of these events was considered a very low probability event, and yet this is the kind of advancement work that long depended on human memory, instinct, timing, and a bit of serendipity.

Now, imagine Rebecca's story within an AI-informed advancement ecosystem. A predictive model flags the article by the alumna and links it to her past giving theme, an integration engine connects that update to the private institutional data, and the forgotten institutional priority surfaces the dormant lab proposal. A natural language system scans the faculty award and connects all the dots, alerting the relationship manager of these connections and recommending a personalized outreach within 48 hours by email. A generative assistant drafts the first message, reminding the relationship manager of the proposed action the next day in case they were on the road or missed the first system notification, pulling the relevant context from donor history, lab proposal, and the faculty award in a note ready for a human to review and send. No lucky timing. No miracles. Just systems talking to each other, continuously and predictably.

DESIGNING FOR SERENDIPITY

AI does not replace intuition; it raises the odds that intuition is activated or triggered, moving from *hoping* for alignment to *engineering* that alignment. In a way, moving from isolated light bulbs to a connected circuit. When the Four Data Frontiers are linked, the chance of a meaningful connection increases exponentially and happens much more frequently.

Five-Step Playbook for Activation

Having a framework is a great starting point, but how do you implement it? To make the Four Data Frontiers actionable, here's a five-step playbook, a sequence of clear and concrete steps that build on one another and hopefully allow you to secure quick wins that demonstrate progress and encourage buy-in:

1. **Map What You Have:** Gather your team for a data mapping session. Whiteboard all the streams of information that your advancement team currently has, and plot them into the four quadrants. At this

point, don't worry about gaps yet; just capture everything and see where it fits. You're likely to identify some blind spots and very likely to have one or two "aha" moments.

2. **Clean and Connect:** Now you can choose the identifier that will link all the records across the different systems. In other words, the key that will let you join all this data. (In most cases, the easiest choice is an email address and not a unique constituent ID, since some of the external datasets are likely to match faster on a common email.)
3. **Pilot with Purpose:** Use a simple pilot now that your data is mapped and connected. You will be inclined to boil the ocean at first by trying to do too much or everything at once. Try to suppress that urge and choose something simple, maybe one cross-quadrant, and create an automated data trigger for it. For instance, configure a rule in your donor information system where *if* an alum updates their job title on LinkedIn to "CEO" or another C-Suite title, and that person is flagged in your donor information system as having a major gift proposal, then notify the relationship manager immediately and mark that prospect for priority outreach. As you develop this pilot, test it, and have the team see its value. You can add more rules and make it more complex *if* the donor makes their first $1,000 or more gift and lives in a region where you're hosting an event. In that case, you might alert your events team to send a personal invite and thank the donor for their generous gift.

4. **Automate and Govern:** With the pilot showing value to your team, start thinking of how to scale and sustain the system. This could mean automating the hygiene part of your data management, minimizing the manual effort in cleaning up the data. Establish clear governance practices to keep things on track, and set up regular data syncs so you can easily push things to and from your donor information system.
5. **Layer in AI and Act:** Add an AI-driven tool to supercharge insight generation and move to action. This is where the excitement really kicks in, where you can do things like build a predictive analytics model to score which alum is the most likely to answer what specific campaign, or which donor opened your last five emails and attended an event in the past three months.

   Systematically doing this will increase trust with your relationship managers, who will be more than likely to spend less time on guesswork, research, and contact reports and spend more time speaking with donors, engaging with alums, and thinking about strategy.

   This five-step playbook can help you methodically move from data chaos to an insight-driven operation. Equally important, each step can help build confidence and culture, with your team seeing the gradual benefit of integration early on in the process. This foundation is critical, for sustaining an AI-ready advancement program is not about AI tools and data, but more about people and culture.

GOVERNANCE IS CULTURE, NOT PAPERWORK OR TOOLS

When people hear "data governance," eyes usually glaze over. The term gives the impression of a thick binder with the word "policy" on it and a long compliance checklist. Within an AI-informed and data-driven advancement office, governance is less about that and more about defining a culture that everyone buys into. In this instance, good governance is a behavior.

At Clark University, that meant defining guiding principles that the team believed in, a two-paragraph section in the Gift Acceptance and Advancement Policy that authorized the advancement team to use AI tools, appointing a committee to oversee the approval of specific AI tools and whether or not they align with institutional values, and a board oversight function from the board advancement committee. Most of that is about conversations and colleagues defining boundaries, and less about pages and pages of policies and binders.

These conversations include basic but fundamental questions: How would this data help provide a better experience for our donors? Do we really need to collect a particular data point?

As a general principle, if we cannot share a specific data point with a donor, we probably should not collect it. Donors and prospective donors at large should never be surprised by what we know about them. The AFP Donor Bill of Rights perfectly portrays this when it states that donor information will be handled with respect and confidentiality.

Restricting data access is another way to preserve donor privacy. Some institutions are instituting practices where only their research team can see the detailed raw data, and the rest of the team can see the synthesized analysis or takeaways from the data. By restricting direct data and focusing more on actionable knowledge and insights, we are likely to reduce the chance of misuse or even bias while still empowering staff to act. It's, in a way, akin to the "need-to-know" security model.

Common Risks and Practical Mitigations

Risk: Over-alerting creates noise and alert fatigue.
Mitigation:

- Rate-limit alerts.
- Bundle signals into a daily digest per relationship manager.
- Only repeatedly highlight tier-one triggers.

Risk: Model bias or spurious correlations.
Mitigation:

- Quarterly precision/recall checks by segment.
- Require reviews by both the researcher and the relationship manager before rule changes.

Risk: Privacy creep with private data.
Mitigation:

- Minimization of data collected is key.
- Only collect what you will use.
- Produce role-based access and sunset rules (i.e., auto-expire certain fields and data points).

Risk: Tool drift without adoption.
Mitigation:

- Provide monthly 30-minute training.
- Refresh the playbook quarterly.
- Provide simple workflows inside your relationship managers' tools.

Sunsetting certain data is also important; data is not wine, and it doesn't usually improve with time. Your institution can institute rules around when certain data should expire and be removed from the primary donor information system (and perhaps be stored in an archived location).

Most nonprofits still lack a comprehensive analytics strategy. Counting on accidents and sparks to get the next life-changing scholarship is not a future-ready strategy. In the nonprofit sector, there's a lot at stake for the people we serve, and taking intentional steps toward future-readiness could amplify our impact.

Chapter Two Recap

Core Message

Fundraising success in the AI era depends on integrating four types of data: private donor data, public donor data, private institutional data, and public institutional data.

Key Takeaways

- AI insights help us move from static reports that we react to weekly, monthly, or quarterly to live insights that show our alumni and donors that we are active, engaged, and aware.
- Insights from donor data and institutional data (from privately and publicly available sources) provide the most reliable donor strategies.
- Culture shapes governance.
- Signals from donor activity (promotion, a priority, an update) become sparks for stronger engagement and larger gifts.

Chapter Endnotes

1. Ketil Gjerstad, Justin Manly, Gerry Hansell, et al., *The Vitality Code: How Growth Leaders Master Strategy, Technology, People, and Culture* (Boston Consulting Group, April 2025), https://web-assets.bcg.com/d7/af/c5359a8447f0aab8985c76a26645/the-vitality-code-apr-2025.pdf.
2. Nidhi Arora, Daniel Ensslen, Lars Fiedler, et al., "The Value of Getting Personalization Right—or Wrong—Is Multiplying," McKinsey & Company, November 12, 2021, https://www.mckinsey.com/capabilities/growth-marketing-and-sales/our-insights/the-value-of-getting-personalization-right-or-wrong-is-multiplying.
3. "Mission," The EverTrue-Clark Advancement AI Lab, https://www.clarku.edu/advancement-lab/.
4. Brent Grinna (founder and president, EverTrue), interview by Joe Manok and David Woodruff, October 2025.
5. Sarah TeDesco, *AI for Nonprofits: How to Harness Its Full Potential*, DonorSearch, February 17, 2026, https://www.donorsearch.net/resources/ai-for-nonprofits/.
6. Adam Conner-Simons, "Using Artificial Intelligence to Improve Early Breast Cancer Detection," *MIT News*, October 16, 2017, https://news.mit.edu/2017/artificial-intelligence-early-breast-cancer-detection-1017.

7. "Driving Breakthroughs in Healthcare Through Collaboration Between the Life Sciences and Technology," Jameel Clinic, https://jclinic.mit.edu/about/.

CHAPTER THREE

Global Models for AI Governance and Ethics

Imagine a scenario where a development director of a small nonprofit works with a donor who wants to give a $5 million gift to her organization. By the time she arrives at her office, she already has 20 emails from colleagues and board members sharing concerns that the donor has been fined $2 million for environmental violations. Within an hour or two, the director most likely gathered colleagues from finance, legal, communications, operations, and at least a board member to discuss the situation and determine how to proceed. Our values and codes of ethics show up in these moments. Advancement professionals and nonprofit leaders have navigated these ethical questions guided by acceptance policies, guiding principles, institutional mission and values, due diligence standards, and board oversight. With every decision they make, they weigh the risks associated with donor generosity against aligning opportunities with integrity.

These same principles and processes now apply to the digital world as AI reshapes personalization, research, segmentation, and analytics, with advancement leaders spending time thinking about applying the same level of rigor they bring to accepting major gifts to the tools that power receiving these gifts. The questions that you can ask yourself here include:

- Who could be harmed, and who benefits from this technology?

- Does automation erode or strengthen our relationships with donors?
- How can we ensure accountability, transparency, and consent when algorithms touch donor data?

In a way, gift acceptance and the adoption of AI share a moral foundation: Both require clarity of purpose, a commitment to stewardship over convenience, and an independence of judgment. The true effort should be put into building the systems, culture, and habits to respond to these ethical dilemmas when they appear and not to predict every one of them. This chapter explores how guiding principles that once helped us govern gift policies (due diligence, proportionality, transparency, and accountability) can now form the foundation of a global model for ethical AI governance in the nonprofit sector. The tools may be new, but the ethics are not.

Ethics, in both cases, is our strategy in action: a way to turn pressure into principled decisions that align with our values. The rest of this chapter translates these questions into maturity models, structures, oversight mechanisms, and operating practices so that innovation can move faster precisely because it's moving safely.

These moments are becoming much more frequent as AI enters the daily operations of nonprofits. We already live in an era where algorithms can guess a donor's capacity and inclination to give, generate talking points, and even write full appeals. What happens when they guess wrong? Or when they reinforce bias?

THE AI ETHICS MATURITY SPECTRUM: A FRAMEWORK FOR READINESS[1]

Ethical AI does not emerge from goodwill and good intentions alone. It requires cultural fluency, structure, and leadership. Our framework, the AI Ethics Maturity Spectrum, defines five distinct levels of ethical AI governance in mission-driven institutions. Each level describes the state not just of policy but of mindset.

Advancement AI Maturity Framework

AI Ethics	Level I: Ad Hoc	Level II: Aware	Level III: Defined	Level IV: Policy-Guided	Level V: Integrated
Ethical Practices	No review; Ad hoc tools	Some bias; privacy talks	Clear but basic Guidelines	Formal policy; trained staff	Embedded; continuous learning
Sample Scenarios	AI appeals sent unreviewed	Should we say this was AI generated?	Requires approval for AI content	Annual audits; vendor checks	Teams co-create; full disclosures
Tools and Oversight	Free tools; no clear checks	Prompt libraries	Board-reviewed workflows	Auditable platforms	Ethics Committee; Board mandate
The ethics of not using AI deserve equal scrutiny -- inaction has consequences too					

Level 1: Ad Hoc

There are no rules or ways to keep an eye on AI use because it's still new. Teams use AI tools on their own, often using free, untested tools like ChatGPT or Grammarly, without the institution being able to see them. Leaders either don't know or don't care, and there's a lot of ethical risk, but it's often hard to see.

Level 2: Aware

Someone needs to point out the need for policies for awareness to happen. You start seeing threads like, "Don't paste donor data into ChatGPT." Ethics is relatively reactive. Leadership is interested in the use of AI but remains hands-off. The nonprofit begins to define what shouldn't happen, but not yet what *should* happen or how.

Level 3: Defined

Policies are written. Teams follow approval workflows for tool adoption with clear oversight or an ethics committee.

Level 4: Policy-Guided

AI governance is organized and watched over. An ethics review is often needed for project intake, and an Intentional Legitimate Interest Assessment (ILIA) comes up. There are logs. Board oversight starts, and ethics becomes a part of the lifecycle.

Level 5: Integrated

Ethics is built in and requires ongoing learning. The nonprofit has regular third-party audits, updates its policies every year, and builds ethical reflexes into hiring, training, and buying tools. Leadership sees AI ethics as a way to stand out from the competition, not as a limit.

Each level moves from reactive to proactive and from policing to enabling. The team would begin to see ethics as a part of daily practice and not red tape meant to slow them down.

Choosing not to use AI, whether out of fear, change, or a lack of understanding of technology, carries its own ethical consequences. The ethics of inaction deserve the same scrutiny as the ethics of action. When we fail to explore doing our job not only responsibly, but also faster, smarter, cheaper, and in a more personalized manner, are we the best stewards of donor funds or public trust?

A TWO-LAYERED OVERSIGHT MODEL[2]

In 2023, before using any AI tools at the Clark University advancement office, we started by going to the board of trustees to formalize a governance structure for the ethical use of AI in advancement. Earlier that year, the vice chair of the advancement committee of the board, Erica Penn, and her husband, Kevin, offered us time with a family friend of theirs who developed advanced AI tools for businesses and helped us better understand some of their basic uses for alumni relations. As tools started to surface, we realized that prior to jumping into the deep waters of AI, we had best develop a clear oversight model. Through a collaboration between our advancement, legal, IT, and president's offices, the university created a section in the Gift Acceptance and Advancement Policy of the board that was both clear and effective. Essentially, every AI tool that touches donor or constituent data must pass through two layers of review.

The strength of this model is how replicable it can be, as it ensures that no single office or individual can approve, deploy, or monitor an AI system in isolation.

Layer 1 - Executive Approval: The Operational Review

Any AI tool or application must be vetted by a cross-functional internal committee that includes representatives from IT, general counsel, advancement, president's office, and, in some instances, academic expertise. The internal review applies standard intake to review purpose, mission alignment, donor privacy, data exchange, human oversight, bias, fairness, transparency, and other relevant elements.

Layer 2 - Board Oversight: The Strategic Review

The second layer elevates accountability at the governance level. A subcommittee of the board of trustees, in our case, the advancement committee of the board, which meets at least three times a year, receives briefings on AI use cases, outcomes, and tools being used. It provides board members with an opportunity to engage with the topic and ask questions and potentially audit these platforms, tools, systems, and algorithms.

GLOBAL GOVERNANCE MODELS: GUARDRAILS THAT ENABLE INNOVATION

AI has forced nearly every sector to ask the same foundational question: How do we innovate responsibly? Governments and industry groups around the world have answered by building frameworks that protect people while allowing data-driven progress to continue. These laws and standards are not bureaucratic barriers but rather ethical guardrails that enable sustainable innovation and responsible adoption. For nonprofits, it's important to remember that the public assumes that we hold ourselves to a higher standard of care. Each framework below offers a blueprint for legal and ethical use of AI.

General Data Protection Regulation (GDPR)[3]

GDPR, a European Union (EU) regulation enacted in 2018, remains the most comprehensive privacy regime worldwide. The spirit of it is simple: Individuals, not institutions, own their data. This regulation provides every EU resident with the explicit right to know what data is held about them, how it can be deleted, and why it's used.

Why does GDPR matter for nonprofits?

- GDPR is relevant to a US university soliciting gifts from an alum in Paris.[4]
- Data can only be processed under specific conditions of legal obligation, consent, vital interest, contract, legitimate interest, or public task. Fundraising typically relies on legitimate interest, and it must be documented through a Legitimate Interest Assessment (ILIA).[5]

- Data minimization, being able to collect only what is necessary, is a key principle.
- The donor must be informed if AI is used to profile the donors. For example, predicting their capacity to give. If the decision has a meaningful effect (say, being included or excluded from campaigns), the individual has the right to a human review.
- Individuals have the right to request the deletion of their data fully unless there's a legal or, in some cases, mission-critical reason to retain it.
- Noncompliance leads to serious penalties, including up to 4 percent of global revenue.
- Many organizations outside Europe adopt the principles voluntarily.

Family Educational Rights and Privacy Act (FERPA)[6]

In the United States, FERPA protects student privacy, such as grades, course enrollment, club membership, and disciplinary history. If such information is used for donor modeling, advancement offices should make sure that, if they're using, say, club membership, it's obtained from a directory that the student didn't opt out of being shared or used with proper consent.

Health Insurance Portability and Accountability Act (HIPAA)[7]

For nonprofits operating in or alongside healthcare, HIPAA defines the ethical perimeter around patient information as it regulates how Protected Health Information (PHI) can be used or shared.

Under HIPAA, even fundraising communications must stay within tight boundaries:

- Permitted use defines what data they can and can't use, mostly around basic demographic data, treating physicians, and a few other details.
- Every fundraising message must include a clear and easy way for recipients to decline future solicitations.
- Development staff must be trained in how to handle PHI data securely.

California Consumer Privacy Act (CCPA) and California Privacy Rights Act (CPRA)[8]

While most US privacy law remains sector-specific, California's CCPA and CPRA mark a turning point toward a national baseline. These acts allow residents of California the right to access, correct, and delete their data and to be able to opt out of its sale or sharing. Even though nonprofits are largely exempt, donors rarely distinguish corporate and charitable obligations and might have the same expectation if their data is mined by a company or their university.

Some analysts note that although nonprofits might not meet the "business" definition of the law, their vendors or data partners might, which means obligations might flow to them indirectly, so it's a good practice to adopt the principles of the law into the nonprofit framework.

AI Risk Management Framework (RMF)

Laws define what you can and can't do. The National Institute of Standards and Technology (NIST) created the US AI RMF, a framework that shows how to deploy correctly.[9] Released in 2023, it has four repeatable steps: govern, map, measure, and manage. These steps help you find, control, and lower AI risk. NIST's framework connects ethics and action for nonprofits. For nonprofit organizations, you can think of these as boxes to check or treat them as building blocks for trust.

AN ECOSYSTEM VIEW: FEDERAL, STATE, AND PRIVATE SECTOR MODELS FOR COORDINATION

No single entity can tackle AI readiness alone; it takes an ecosystem and an ecosystem view even within your organization. Around the world, governance models are emerging at multiple levels, each creating an "AI ecosystem" suited to its context. This section examines three lenses of ecosystem strategy: national, state, and private sector.

National Lens: America's AI Action Plan (2025)[10]

At the national level, the United States has framed its AI efforts as a broad ecosystem race. This plan positions AI development as a strategic imperative for economic and national security. "Whoever has the largest AI ecosystem will set global AI standards and reap broad benefits," the plan declares, likening the challenge to a new space race.

State Lens: Massachusetts's AI Strategic Task Force (2024)[11]

At the state level, Massachusetts has been among the first to use a cross-sector ecosystem approach to AI. The commonwealth set up a 25-member AI Task Force in 2024 to investigate the effects of generative AI and make the state a leader. The state's economic development and technology secretaries co-chaired this task force, which brought together people from academia, business, government, and civil society.

The mission was twofold: to encourage leading industries in the state to adopt AI to help drive economic growth and efficiency, and to ensure the state coordinates efforts to attract AI talent and investment. This is a holistic ecosystem view recognizing that to be an AI state was to link universities' start of large enterprises on a public factor network. The task force launched a public-private AI hub backed by a proposed $100 million fund for AI initiatives. The key difference from the federal approach is that the state model has an emphasis on regional strengths and inclusive growth, stressing workforce development, education, and ethical guardrails.

Private Sector Lens: DAIMLAS: Building Local AI Ecosystems

The private sector ecosystem building has become a business model. DAIMLAS is a notable company that turned itself into an AI ecosystem builder.[12] It works with communities, governments, and organizations to build, manage, and grow local assistance at the national or state levels. As their chief AI ecosystem officer, Joseph Wehbe, says, the company "maps out everything" from people to academic programs, student clubs, funding opportunities, labs, government agencies, startups, and corporate players.

The key lessons for nonprofits are that an AI ecosystem can take many forms:

- providing overarching direction, funding, and norms;
- mobilizing regional and local strengths, and partnerships with communities; and
- bringing together networking and speed by bringing together people, platforms, funding, and organizations.

What unites these views is that coordination beats isolation, and that ecosystem thinking is readiness thinking.

Expert Voice: Nathan Chappell

Responsible and Beneficial AI Framework by Fundraising.AI[13]

The most credible and comprehensive signal that the nonprofit AI ecosystem moved from theory to practice came in 2023. After a conversation that Nathan Chappell led informally at the Association of Fundraising Professionals Conference, it resulted in the birth of Fundraising.AI and introduced the Responsible and Beneficial AI Framework. When he introduced the framework, the nonprofit sector witnessed the beginning of a shift that was long overdue. For years, fundraising has been adopting a few elements of AI here and there, from internal tools to vendors doing basic levels of analytics, and a great consolidation with the creation of Gravyty by Adam Martel. The public announcement by OpenAI in September 2023 transformed everything and made systems, processes, and even platforms near obsolete.[14]

Chappell's framework, launched through the newly formed Fundraising.AI coalition, filled that vacuum. It represented not another vendor or product or solution, but a shared code of conduct that is seen by some as an ethical north star for the industry.

Chappell, who now leads the AI efforts at Virtuous, posits that the fundraising economy is built on trust and represents the half-trillion-dollar-a-year system. AI cannot merely make that system faster, but it must make it fairer, safer, and more transparent. The framework's power is not in a new slogan, but in creating a shared baseline for how the sector builds and governs intelligent tools. It asks leaders to treat donor data as a stewardship obligation and not as raw material. It demands that responsibility be engineered rather than assumed when designing models, contracts, review processes, and culture.

For years, AI crept into advancement through wealth screening, propensity scoring, content generation with uneven disclosure, and no common vocabulary for what "responsible" meant. Fundraising. AI, with its advisory board of mostly corporate leaders, changed that center of gravity. It articulated a set of principles around privacy and security, inclusiveness and fairness, transparency and explainability, and accountability and oversight. It insisted that they be present across the entire lifecycle of an AI system, from purpose definition and data collection to testing, deployment, and retirement. In practice, that means building for the minimum necessary use of data, documenting how and why models reached a certain conclusion, and keeping humans in the loop so that predictions inform rather than replace judgment. It also means giving supporters meaningful choices, making it easy to opt out of analytics uses, and ensuring that any third-party vendor touching donor data meets the standards you claim for yourself.

PROFILING AND PREDICTIVE ANALYTICS: DOING IT RIGHT

Few areas of AI stir as much tension in nonprofit work today as profiling. In an advancement setting, efficiency must meet empathy. When done well, predictive analytics can make fundraising smarter, more equitable, and more personalized. When done poorly, it can alienate the very donors and communities the nonprofits seek to engage.

Profiling is the practice of using data to make informed guesses about people's behavior, such as who is likely to give, volunteer, attend an event, or accept a meeting request. It's what allows AI to recommend the best next action. The power is undeniable, but the danger lies in forgetting that these actions and predictions are about people and not what movie to watch next on Netflix or what to buy on Amazon.

The Promise: Smarter, More Inclusive Fundraising

Done ethically, predictive analytics can level the playing field and, in a way, democratize fundraising. By analyzing past interactions, demographic data, and engagement patterns, nonprofits can surface new prospective donors who might otherwise be overlooked. For example, one university advancement team discovered through predictive modeling that alumni who engaged heavily with faculty on LinkedIn but had no prior giving history were as likely to donate as mid-tier donors with a decade of giving history. This helped them expand, not narrow, their prospective donor pool.

The risk is when predictions replace and don't augment people. The nonprofit world has already started to witness this. In 2017, the UK Information Commissioner's Office fined several major charities for secretly using wealth screening and data enrichment services without informing donors. Although the fines were modest, the reputational damage was severe.[15]

The AI Paradox: When an AI Flags Its Own Creator

At a conference in the summer of 2025, we were introduced to a new generative AI tool, and we decided to ask the tool itself if we should partner with its creator, the parent company. The AI model came back with a confident, "No. Wait at least 12 to 18 months before considering a partnership." The model highlighted ethical practices that it flagged for us, as well as legal battles lining up and a whistleblower report in the media. This created a paradox; on one hand, this is the best model on the market during the writing of this book, and on the other hand, it advised against collaboration. We didn't collaborate.

Principles for Ethical Profiling

The difference between responsible use of AI and reckless profiling is not sophistication, but governance. Ethical profiling starts with a clear question: Just because we can predict something, should we do so?

Below are six practices that separate ethical profiling from potential exploitation.

1. **Conducting an Institutional Legitimate Interest Assessment (ILIA)**

   An ILIA is the equivalent of a preflight checklist. Before starting with any analytics projects, the ILIA forces teams to consider necessity, proportionality, and fairness. The process is simple but profound, asking:

   - What is the purpose of the profiling? How does it advance the mission of the organization?
   - Is there legitimate interest or consent for processing the data (a lawful basis)?
   - Is the impact on the individual proportionate to the benefit for the organization?

- Would the individual reasonably expect this kind of analysis?
- Are there any human oversights and safeguards we will put in place for data minimization, opt-outs, and bias testing?

2. **Offering Opt-Outs and Choices**
   Opt-outs and, in some instances, opt-ins, should be offered to all constituents, giving them the ability to say no to their data being mined, analyzed, or even stored. The principle might be cumbersome, but it communicates values, and in some jurisdictions, it's a legal requirement.
3. **Keeping Humans in the Loop**
   No AI system should ever make any consequential fundraising decision alone. Imagine an AI model flagging a long-time donor as "low potential" because they didn't donate last fiscal year and then losing that donor due to that oversight. This is why ethical AI is augmented intelligence and not autonomous. Although fully autonomous AI tools are on the rise, we consider AI as a decision-support tool and not a decision-maker, a principle that stems from the fundamental assumption that people give to people. Human connection is a key element to keeping philanthropy in the relational space and not in the transactional space.
4. **Making the Logic Explainable**
   Donors and staff alike should be able to easily understand in plain terms how an AI model is being used or reaches its conclusion. Transparency is the antidote to mistrust.

5. **Practicing Data Minimization**
   Ethical analytics collect less, not more. The principle, borrowed from HIPAA, applies to all AI use. Every data field should have a purpose tied to a mission.
6. **Audit for Bias and Equity**
   Every predictive model should be reviewed for fairness across key demographics. These bias audits are the ethical equivalent of performance testing.

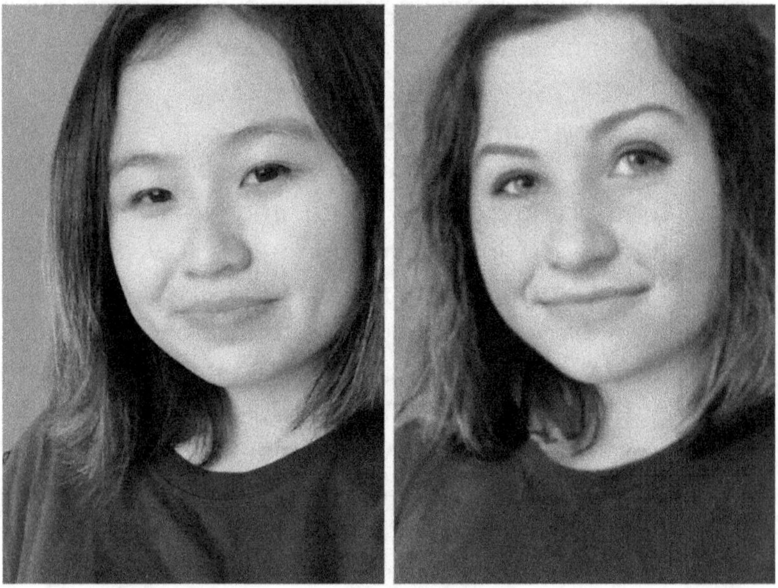

Take the case of Rona Wang. In July 2023, the *Boston Globe* published a story highlighting racial bias.[16] Rona used an image generator to create a professional LinkedIn profile, and the AI tool decided to change her facial features, making her look Caucasian, with rounded, blue eyes and a lighter skin color.

The bias was not clear until this specific query was made, and the output visibly showed an issue. As you think of any model or dataset that you are working on, think of the potential blind spots and the invisible bias. Include colleagues who better understand the data and its use to make sure you're as inclusive as you can be with the algorithms, models, products, and services that you are using.

When Profiling Crosses the Line

Even the most well-intentioned analytics can go astray, but there are clear warning signs that can help us monitor that:

- Are you becoming more focused on efficiency than on your mission?
- Are you drifting away from being relationship-driven and becoming data-driven instead of data-informed?
- Are your predictive scores becoming predictive quotas?

The Payoff of Responsible Personalization

When we do it right, predictive analytics doesn't just increase our efficiency; it also increases trust, augmenting our sensemaking capabilities with information most relevant to our strategy. Donors see more communications that reflect their interests and values, rather than generic appeals.

Staff focus their time on building relationships and spend less time staring at Excel files. When done correctly, ethical use of predictive analytics reframes AI from being a black box to becoming more of a mirror. It reflects to the donors the same respect that we expect from them. In doing so, the data transforms from being a compliance risk to an asset in building and growing human relationships.

The question for leaders is not whether to use predictive analytics, but more so *how* to use it wisely and safely. One way to explore this is by using a model by technologist Harlan Harris from his essay "Transparency, Trust, and Proprietary Predictive Analytics."[17] Dr. Harlan explores how predictive models could erode public trust when they operate like black boxes and argues that the most important ethical challenge is transparency.

It's crucial for those affected by a model and AI to understand it, at least in broad terms: how it works, what data it uses, and the outcomes it drives. That makes trust a function of performance as well as of explainability. Most newer versions of generative AI models are trying to explain to the users what they are using to answer our prompts. They started sharing what they are doing, the sources being used, and references to the specific elements of their research. Explanative AI is, therefore, a key element of public trust. Harris acknowledges that it is especially difficult to explain models that are proprietary, such as internal algorithms or commercial models, and offers a potential solution by creating a shared accountability model between developers, end users, and other stakeholders.

THE ETHICAL SOFTWARE OF YOUR NONPROFIT: INTERNAL AI GUARDRAILS

Nonprofits should metaphorically use an ethical operating system. The underlying set of tools and processes helps translate the missions and values into daily use of technology in general, particularly AI. Key components can include:

1. **Personal Position Statement**
   Start with your personal position statement. Highlight your personal comfort and your boundaries of what you think is ethical and unethical use of this technology.

2. **Ethics Charter**
   Create an internal document that defines the institution's values around AI, as well as its commitment to the community and donors. This should cover concepts like the institutional approach to privacy and accountability.
3. **AI Use Policy**
   Define who approves tools and who provides oversight.
4. **ILIA Workflow**
   Use an Intentional Legitimate Interest Assessment (ILIA) tool to evaluate everything from simple tools to the most advanced ones systematically.
5. **Model Cards**
   Create a card for every AI tool deployed, documenting its purposes, inputs used, update frequency, risks, limitations, and escalation path.
6. **Team Training**
   Commit to continuous training as you learn more about this technology and as it continues to evolve.
7. **Incident Reporting**
   Establish a standard procedure to document, record, and report when something goes wrong or isn't in line with its intended purpose.

Case Study: Iyad Rahwan and "The Moral Machine"[18]

Both of us worked at MIT in 2016 when Iyad Rahwan taught there and decided to launch his online experiment called "The Moral Machine."[19] His experiment became the most revealing global ethics experiment in the history of AI so far, asking a simple and extremely powerful question around autonomous vehicles:

In a crash scenario, who should a self-driving car save?

The choices available were left open: A pregnant woman or a doctor? Two elderly people or five jaywalking teenagers? The car's passenger or a group of strangers?

Over 2 million people from more than 200 countries participated, and the results revealed that the answer was neither straightforward nor universal. The data showed three dominant moral clusters: Western, Eastern, and Southern clusters, each with distinct ethical reflexes that reflected centuries of philosophical, religious, and social conditioning. This is vital not just for engineers but for nonprofit leaders whose missions, goals, and constituents cross cultures.

Western Cluster (United States, France, Western Europe)

Participants from this cluster generally prioritize minimizing harm and maximizing benefit. In a sense, they prioritize utilitarian outcomes, favoring humans over animals and sparing younger individuals over their elders. Their moral framework is grounded in individual rights and rationalism, with "fairness before rules" being the conceptual foundation. While participants value law-abiding pedestrians, they make exceptions to reduce harmful results. Ethics are more situational, and "breaking a rule to do good" is acceptable.

Eastern Cluster (China, Japan, Islamic Societies, Singapore, South Korea)

Maintaining reverence for age and wisdom is a distinct characteristic of these respondents. Influenced by Confucian and Islamic traditions, they prioritize social order over the autonomy of the individual, favoring law-abiding pedestrians over jaywalkers.

Southern Cluster (Latin America, Francophone Africa, Parts of the Middle East)

Communal and survivalist ethics were the dominant characteristics of these respondents. They are more likely to spare athletes and working individuals. This reflects the outlook of communities where opportunity might be scarce, and inequality shapes moral perception. At face value, this might appear biased, but with fragile institutions and high economic instability, ethics might shift more towards preserving the community. One Latin American researcher observed, "Morality here often mirrors scarcity." Inequality doesn't just distort economics; it can also shape our sense of justice."[20]

If you manufacture a car in Germany using technology from the US and Asia and then sell it in Latin America, should you program it to adapt to the local culture or to remain in line with standards? Who sets these standards? Rahwan spoke against exporting moral values across different cultures, but believes we should find solutions to these complex problems.

What does any of this have to do with nonprofits or our book? The implications of Rahwan's work are profound for the nonprofit sector. The expectations of what is right, what is acceptable, and what is fair clearly shift across cultures. When global organizations team up with partners, donors, and talent to design decision-making AI systems, they inherit those systems' codes, preferences, and biases.

In Western philanthropy, we see donors wanting transparency, more data, and to see that their support is making a real impact to achieve the "greater good." If AI models were designed for these environments, they would enhance scale, reach, and ROI, putting more weight on metrics like donor conversion and the cost of engagement.

Giving in Eastern cultures would put more value on legacy, community responsibility, and honor. So, algorithms or AI models that feel impersonal would not work well.

In Southern or emerging economies, philanthropy is perceived as a means to an end of correcting a social imbalance, since the organizations and communities being supported might be in much more resource-constrained environments.

Just like we localize language and messages, as global nonprofits work with global donors to make an impact across the world, their AI should also be able to understand how to localize ethics while staying true to the global values of their organizations. Iyad Rahwan's Moral Machine reminds us that the "right thing to do" depends not just on logic, but on the lens through which we view humanity. For AI to serve missions faithfully, it must not be designed as a universal moral authority, but rather as a way to mirror the values of the people that it represents and serves.

The ethics of using certain tools is clearly influenced by people and geography, as evident above by the Moral Machine, but it's also influenced by location, be it physical proximity of the donor and beneficiary, the type of organization that you serve, or the industry you represent.

In 2010, Bashshar Haydar, an ethicist from the American University of Beirut, explored how physical proximity influences our moral decisions, including philanthropy.[21] He shares how, if asked for $10,000 by a random stranger to support another stranger who needs help on another continent, we often think this is someone else's problem or that billions of other human beings can help that person, and it doesn't have to be you. He presents the same stranger drowning near a bridge, and you're the only person driving nearby. The only way to save their life is to drive in their direction, resulting in tens of thousands of dollars in personal property damage. Many of us would do it without hesitating.

Proximity and location are major influences on what's considered ethical. If you're working at a healthcare organization, move to a small liberal arts college, and then to a large global nonprofit based in Switzerland, what your community is comfortable with doing and what it considers ethical could vary. That's why it's important to highlight your personal position statement to remind yourself what you're comfortable with and what your personal values are, irrespective of tools, processes, and institutions or geographies you're applying these values and ethical principles to.

The Moral Machine reminds us that algorithms reflect not only data but values. Those values must be reviewed, evaluated, and chosen, not assumed.

Three leadership commitments can help you turn ethics into strategy.

1. **Normalize transparency**, and publish your AI use cases and learnings as trust grows with visibility.
2. **Empower experimentation** with guardrails to encourage innovation, but within documented, reviewable boundaries.
3. **Institutionalize reflection** from planning premortems and postmortems on AI projects as routine as financial audits. Ethics is a muscle, and you strengthen it through repetition.

In a data-rich world, treat ethics as you would your brand promise, your most valuable currency, along with trust.

Chapter Three Recap

Core Message

Ethical maturity in the AI era doesn't come from binders and policies, but from translating values into daily decisions, where the most successful organizations would operationalize ethics through action. True governance isn't about slowing the AI innovation cycle, but about building the trust and accountability needed to make that innovation sustainable and well adopted.

Key Takeaways

- Ethics is strategy in action. AI governance succeeds when key principles are embedded in workflows.
- Collect and use only the data you need to achieve mission goals. Less is more in some instances.
- Transform compliance into stewardship, and focus on answering these questions: What data should we collect? Why are we collecting it? And how should we use it?
- Ethical oversight is not static and should evolve continuously.
- Always look at ethical blind spots before they turn into breaches.
- Responsible AI use should not count on the goodwill of the individuals implementing these AI tools, but rather be built into systems, contracts, team training, and board oversight.

Chapter Endnotes

1. Joe Manok, "What Is Your Next AI Move?," *Forbes*, September 29, 2025, https://www.forbes.com/councils/forbestechcouncil/2025/09/29/what-is-your-next-ai-move/.
2. "Use of Artificial Intelligence in University Advancement," in *Clark University Gift Acceptance and Advancement Policy* (Clark University, 2024), https://www.clarku.edu/policies/wp-content/uploads/sites/56/Gift-Acceptance-and-Advancement-Policy_Final.pdf.
3. European Parliament and Council, "Regulation (EU) 2016/679 of 27 April 2016," *Official Journal of the European Union* 59, no. L119 (2016): https://eur-lex.europa.eu/legal-content/EN/TXT/PDF/?uri=CELEX:32016R0679.
4. "Guidelines 3/2018 on the Territorial Scope of the GDPR (Article 3)," European Data Protection Board, November 12, 2019, https://www.edpb.europa.eu/sites/default/files/files/file1/edpb_guidelines_3_2018_territorial_scope_after_public_consultation_en_1.pdf.
5. "How Do We Apply Legitimate Interests in Practice?," Information Commissioner's Office, accessed March 1, 2026, https://ico.org.uk/for-organisations/uk-gdpr-guidance-and-resources/lawful-basis/legitimate-interests/how-do-we-apply-legitimate-interests-in-practice/.
6. "What Is FERPA?," Protecting Student Privacy: U.S. Department of Education, accessed March 1, 2026, https://studentprivacy.ed.gov/faq/what-ferpa.

7. "Summary of the HIPAA Privacy Rule," U.S. Department of Health and Human Services, March 14, 2025, https://www.hhs.gov/hipaa/for-professionals/privacy/laws-regulations/index.html.
8. "California Consumer Privacy Act (CCPA)," California Department of Justice, March 13, 2024, https://oag.ca.gov/privacy/ccpa.
9. "Artificial Intelligence Risk Management Framework (AI RMF 1.0)," National Institute of Standards and Technology, January 2023, https://nvlpubs.nist.gov/nistpubs/ai/NIST.AI.100-1.pdf.
10. *Winning the Race: America's AI Action Plan* (The White House, 2025), https://www.whitehouse.gov/wp-content/uploads/2025/07/Americas-AI-Action-Plan.pdf.
11. "Executive Order No. 629: Establishing an Artificial Intelligence Strategic Task Force," Office of the Governor, Commonwealth of Massachusetts, February 15, 2024, https://digitalgovernmenthub.org/wp-content/uploads/2024/02/EO-629.pdf.
12. "DAIMLAS #1 Ranked Artificial Intelligence Ecosystem Builder Expert Platform," DAIMLAS, accessed March 1, 2026, https://daimlas.com/.
13. "Fundraising.AI Advocates Responsible and Beneficial AI for Fundraisers," Fundraising.AI, accessed March 1, 2026, https://fundraising.ai/about/.
14. "ChatGPT Can Now See, Hear, and Speak," OpenAI, September 25, 2023, https://openai.com/index/chatgpt-can-now-see-hear-and-speak/.

15. Owen Bowcott, "UK Charities Fined by Watchdog for Wealth Screening of Donors," *The Guardian*, April 5, 2017, https://www.theguardian.com/uk-news/2017/apr/05/uk-charities-fined-by-watchdog-for-wealth-screening-of-donors.
16. Spencer Buell, "An MIT Student Asked AI to Make Her Headshot More 'Professional.' It Gave Her Lighter Skin and Blue Eyes," *The Boston Globe*, July 21, 2023, https://www.bostonglobe.com/2023/07/19/business/an-mit-student-asked-ai-make-her-headshot-more-professional-it-gave-her-lighter-skin-blue-eyes/.
17. Harlan Harris, "Transparency, Trust, and Proprietary Predictive Analytics," Medium, February 27, 2017, https://medium.com/@HarlanH/transparency-trust-and-proprietary-predictive-analytics-e4155030c55f.
18. Iyad Rahwan (Max Planck Institute for Human Development, Center for Humans and Machines), interview by Joe Manok and David Woodruff, 2025.
19. Edmond Awad, Sohan Dsouza, Richard Kim, et al., "The Moral Machine Experiment: 40 Million Decisions and the Path to Universal Machine Ethics," *Nature* 563, no. 7729 (2018): 59–64, https://dspace.mit.edu/bitstream/handle/1721.1/125065/Moral%20Machine%20Paper.pdf.
20. Awad, et al., "The Moral Machine Experiment."
21. Bashshar Haydar (professor, American University of Beirut), interview by Joe Manok and David Woodruff, 2025.

CHAPTER FOUR

From Disruptive Innovation to Purposeful Design: Frameworks for Responsible Change Management

To some, fundraising may seem like an endeavor that has experienced little change over the decades. It has been shaped by unshakable operating principles, such as moves management, campaign planning, and case for support generation. If you take a closer look, it becomes clear that even these foundational practices have evolved significantly, thanks to technological advances, cultural shifts, and a host of process innovations that have ushered in greater efficiency and monumental growth in revenue. In their annual report, Giving USA totaled American 2024 donations at nearly $600 billion.[1] Even as a cumulative report, nonprofits nationwide could not scale to such heights without undergoing enormous and constant change, driven by disruptive innovation and purposeful design.

DISRUPTIVE INNOVATION AND PURPOSEFUL DESIGN IN FUNDRAISING

Disruptive innovation was coined by Clayton Christensen in his seminal work in the 1990s.[2] This describes when a new entrant enters the market, challenging established businesses as it continues to move up-market.[3] This process takes time and spans several steps that eventually lead to disruption, becoming the more profitable, mainstream adoption. Disruptive innovation for fundraising can come in many forms, including a shift to digital giving, the emergence of crowdfunding platforms (such as Kickstarter and GoFundMe), social media and viral communications, and mobile giving. While considerable, these disruptors don't compare to the new force in fundraising: AI. With it, we see the beginnings of autonomous fundraising, prospect research, and stewardship functions.

Design follows innovations. It is how teams can make these AI "disruptors" work for them. This approach is anchored in intention and clarity, moving more quietly and gradually than disruption, with an endgame of careful iterative improvement. Purposeful design begins by asking who we are serving, what matters most, and how we build with integrity. It produces solutions that most often build on success and common sense. Examples of purposeful design in fundraising include moves management, personalized stewardship, and mission-aligned branding. Great design is not about novelty; rather, it is about understanding context and building trust so that enduring and reliable processes can result.

FUNDRAISING IN A FIELD OF CONSTANT CHANGE: THE AI PROMISE

The promise of AI is everywhere, and so is the confusion. When change happens, our instinct is to resist it, embrace it, or just freeze. This is not unique to changes driven by technology, but any kind of organizational change or cultural transformation. We are comforted by familiarity, so we often fear the future and the unknown. This is crucial for fundraising success; we are more likely to embrace changes when we understand the context and purpose related to those changes. With respect to AI adoption, team members need that clarity, or they become directionless, less aligned with strategy, and more ad hoc. In such cases, for example, team members may launch pilots without mandates or guardrails and risk wasting valuable time and resources. AI must be implemented with clear mandates and objectives for it to be accepted. The principles of purposeful design offer a framework for AI adoption that results in more strategic mastery of its powers and intentions. The challenge is in building the organization's capacity to utilize purposeful design fast enough to temper the disruptive changes AI presents.

In this chapter, we will discuss how to turn this commonly chaotic early adoption into a strategic, ethical, and intentional transformation. We will complete the trilogy of data, ethics, and design by shifting the focus from reactive experimentation to purposeful design.

The imperative of AI transformation is clear. If leaders don't design their AI strategy, the AI market will design it for them. The market is flooded with products that represent AI capabilities. Leaders must resist the temptation to turn to the trendiest tool of GenAI in 2024 or Agentic AI in 2025 or to hastily respond to potential pressures from their board or colleagues to do so. Instead, focus on building a trustworthy architecture and governance system that can harness those tools responsibly. The 2025 MIT report on enterprise AI, in which 95 percent of organizations have seen no ROI from their AI pilots, is a stark warning that we should focus on team adoption, workflows, and most importantly, strategic design.[4]

Our aim in this chapter is to ensure readers are among the 5 percent harnessing AI pilots into sustainable and meaningful impact by leading with purposeful design.

Design starts with a hypothesis. Just as a doctor cannot prescribe treatment without a diagnosis, a nonprofit organization should not plunge into AI without evaluating where the need is and how to start. In chapter one, we introduced the AI Readiness State (AIRS) model, a quick self-assessment of your organizational AI maturity. By now, you may have a sense of where your team falls on the spectrum, individually or as an overall unit, from Level I Resistant (skeptical with little or no AI use) up to Level IV Future Ready (where AI is embedded in strategy and operations). Regardless of your AIRS level, this chapter offers crucial insights into purposeful design. Think of it like this: You could be operating at level 3 or 4, but adding tools as you come across them, much like flower picking versus a thoughtfully arranged AI bouquet. You get the most out of your tools and systems if they're a coordinated, intentional effort.

Expert Voice: Christopher Cannon

Why Design Must Precede Automation[5]

Christopher Cannon has spent two decades helping nonprofits and universities modernize their fundraising operations, first with BWF, later with the Zuri Group, and now as the founder of the Generosity Collaborative, Inc. When asked about the impact of AI and automation on philanthropy, he offers a caution that cuts through the hype: "Don't automate chaos." Cannon's point is critical and very simple. Automation amplifies whatever system already exists (clarity or confusion). If your data is fragmented, workflows unclear, and governance undefined, AI will not fix those problems but multiply them at machine speed. He has seen organizations invest in automation before mapping their processes only to discover that they have digitized bottlenecks and made bad habits institutionalized, which he and others refer to as "a Legacy 2.0 mistake." In his words: "You need the fundamentals before you get fancy."

"Automation is a force multiplier," said Cannon. He advised starting with documenting how information flows through your system and where human judgment still adds value to your organization. The goal is to design for clarity first and then for efficiency.

Cannon's insight links directly to the core message of this chapter: design before deployment. Technology should serve a well-defined purpose and not ignore structural or cultural gaps.

Instead, Cannon is seeing too many shops use the "ready, fire, aim" method, assuming that buying equates to implementing.

In this chapter, we will discuss how to turn this natural, chaotic early adoption into a strategic, ethical, and intentional transformation. We will complete the trilogy of Data, Ethics, and Design by shifting the focus from reactive experimentation to proactive design.

The direction of AI transformation seems to be clear. If leaders don't design their AI strategy, the AI disruption will design it for them and their team. That is, poorly managed transformations will diminish trust and risk security breaches and other undesirable outcomes.

Concurrently, the well-managed transformations by other "competitors" in the market may encroach on donor acquisition, retention, and overall engagement. At the same time, you should get out in front of your team's AI transformation, try to resist the instinct to turn to the trendiest tool of GenAI in 2024 or Agentic AI in 2025, or respond to potential pressures from your board or colleagues to do so. Instead, focus on building a trustworthy architecture and governance system that can harness those tools responsibly. Looking back at the MIT report on enterprise AI, where 95 percent of organizations see no ROI from their AI pilots, we should focus on team adoption, workflows, and, most importantly, designing our strategy. The 95 percent that missed the mark were generally characterized in the study as top-down leadership demands rather than organic, client-driven solutions. In this way, that study affirmed the importance of putting design first and getting to the "why" of the effort before building the solution. Our aim in this chapter is to make sure most of us are among the 5 percent turning AI pilots into sustainable and meaningful impact by leading with design.

Design starts with a diagnosis. Just like a doctor can't prescribe treatment without a diagnosis, a nonprofit organization shouldn't plunge into AI without evaluating where it is and how to start.

STRATEGIC FRAMEWORK AND PILOT DESIGN

Implementing a desired strategic framework begins with identifying a high-impact area where AI can advance your mission, solving a real problem, clearly defining success and setting criteria, building a cross-functional team, running a controlled experiment, and securing strong executive sponsorship. From an innovation standpoint, it is like moving from a hackathon mentality to a pilot-as-prototype mindset. Your end goal is to not only determine if a particular pilot or tool is going to work, but to be able to design a blueprint for how AI will operate across your nonprofit if the pilot succeeds.

The ideal starting place is the "mission before pain points." This helps ground you in what matters most. After you study where AI could help amplify your impact, consider your most common pain points and explore how AI can alleviate a persistent challenge. Is it in program delivery, insights, donor personalization, operational efficiency, or constituent outreach? For example, let's say you are launching a new initiative, and your fundraisers are struggling to identify the right prospective donors for these initiatives. In such a case, an AI pilot can focus on identifying what data points can be used to produce this outcome, drawing on the data frontiers from chapter two. Crucially, the pilot idea must align and be filtered through purpose: Does this AI tool align with our values, and would we be proud to scale it if it works? And more importantly, would our donors and prospective donors agree? If the answer to both questions is yes, that is a signal to proceed.

The next step is to define success and risk upfront. What will you consider a win, a percentage increase in engagement? Reduction in staff time on a task? An improvement in service quality? Better response from a donor satisfaction survey? Define the key performance indicators (KPIs) before you start, so you can track and measure success or failure and learn from the experiment. Equally important, however, is outlining the failure criteria and risks associated with the pilot: How would you know if the pilot is causing ethical or reputational risks? When would you slow down, pause, or pull the plug on the pilot?

This is where the Ethics Maturity Model comes into play. During pilot design and execution, an ethical review is required. If you don't yet have an AI review committee, use your pilot to establish one. This will clearly slow down your process timeline, but it will jump-start the prerequisite of your larger AI adoption strategy. In the end, this is a net gain for your program to start with ethics and accountability already at the heart of your work, as opposed to thinking of them as a step or part of the implementation process.

With this strong foundation, assemble a cross-functional team to run the pilot, preferably with some AI skeptics. AI in nonprofits is a team sport and involves data, IT, leadership, program staff, and often external partners.

Include end users like frontline fundraisers and alumni engagement professionals, volunteers, data/tech experts, one or two AI experts, and someone to consider the compliance and legal elements. Encourage the team to operate in an agile mindset with a mandate to surface issues candidly, conduct frequent check-ins, and ensure open communications. Make sure they have air cover to do so with a clear sponsor at the executive level who publicly supports the project and shields it from knee-jerk rejection at the first hiccup.

This touches on the political lens of change: understanding power dynamics and ensuring you have the right allies. Departments are more likely to cooperate with a pilot that is endorsed or sponsored by both the head of advancement and the chief information officer (CIO). If the board is briefed and excited, that excitement will flow to others. Strategic design is about stakeholders and managing relationships and expectations as much as it is about the technology itself and the specifications.

In order to maximize the learning aspect of the pilot, design it for a short timeframe (around, say, three to six months) with phases incorporated in it, like a discovery phase (to research, gather data, and benchmark with peers), a buildout and testing phase, and an evaluation or assessment phase. Encourage the team to scan during discovery, scanning for similar use cases, guidelines, or comparable problems in other industries. This ensures that your design is not happening in an echo chamber, and your team is not reinventing the wheel. The build phase is usually the toughest as most people want to do more (or less), while you want the focus to remain on the pilot's scope. Remind your team that it is important to solve a clearly defined small problem fully, rather than a big problem halfway.

Document everything from your data sources, model parameters, what did not work, decision rules, what worked more than you expected, and any team dynamics that you learned as part of the design. This creates a repeatable process. Be brutally honest with the evaluation: Did the AI actually improve outcomes, and did it do so in a way that is consistent with your values and initial goals and objectives? If a pilot recommendation system resulted in an increase of major giving by five, but introduced privacy concerns or donor discomfort, those cultural signals are as important (if not more) than the additional dollars raised.

This reminds us to pay attention to norms, narratives inside the organization, and fears. Did people feel disempowered? Did the staff morale go down due to working with AI? Did the team celebrate the pilot's outcome, including documenting failures, thereby solidifying a culture of experimentation, or were they just happy it was done? Each test is a sort of cultural experiment that is beyond technology or direct outcomes reported.

Finally, plan for the hand-off or scale-up of the pilot in the design process. The last mile of adoption is also one of the most ignored steps in AI implementation. Are you integrating team learning to facilitate adoption, or is everything going to be shelved after the demo? Although a common tool is to include end users in the process as eventual process owners, sometimes this is difficult with AI, as end-user readiness may vary from design readiness. So you might have to work on this in parallel. This anticipatory planning reflects a strategic design lens where you work toward aligning processes and structures to support the change.

Put simply, you are treating your small pilot as the microcosm of your AI future. If you do this right, which we did not do in the first few instances ourselves, you create not just a proof-of-concept but a template that others in your nonprofit (or sometimes the industry) can follow.

During the 2018 cohort of the MIT Leader to Leader program, five fellows studied how change happens at MIT, what drives change, and what restrains it.[6] Using the Three Lenses of Organizational Change, they examined a strategic lens (actions that unfold through planning), a political lens (actions shaped by power), and a cultural lens (actions that emerge through habit). After interviewing a couple dozen leaders and past leaders from MIT and other institutions, what they found was that change accelerates when leaders switch lenses intentionally, rather than relying on vision, structure, or charisma alone.[7] They also concluded that political coalition-building is perhaps the most underused but most essential skill for sustaining change and innovation in universities and that, through small, daily micro-actions, rather than grand statements, leaders can change cultures that in turn help change organizations.

Effective change is designed, not declared. In an AI environment, this means AI governance is about the culture of the place, adoption is about what can politically happen, and readiness is a combination of both.

DESIGN WITH A FRAMEWORK

The d.school at Stanford University popularized a five-phase design thinking model: empathize, define, ideate, prototype, and test.[9] This can be easily translated to the context of designing AI initiatives in a nonprofit or a university. Each phase offers a repeatable method to ensure that the adoption of AI is human-centered and iterative rather than a traditional linear rollout with a clear beginning and end.

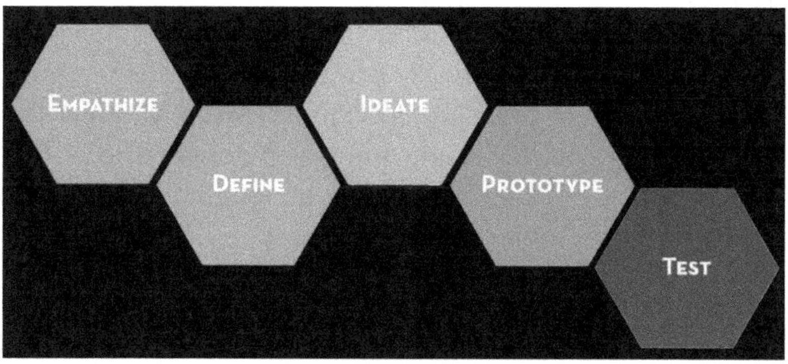

Stanford d.school Design Thinking Model[8]

Empathize: Ground AI in Human Voices

Empathizing with the human voices in a nonprofit advancement context means including frontline fundraisers, donor relations professionals, donors, alumni, and other people who will benefit from, interact with, or be affected by the AI tool or system you are designing. The goal is to map pain points, expectations, and hopes before jumping to tech solutions. You might hold a listening tour, surveys, and workshops. Examine repetitive tasks, where you fall short, or where you spend a lot of time.

Define: Articulate the Problem That Your AI System Will Solve

Your next phase is to turn what you heard into a clear problem statement by synthesizing the empathy findings to pinpoint the core needs. While doing so, it is important to focus on the human need and not a technical issue. Not solving for the right problem could delay adoption or lose it altogether.

Ideate: Generate Responsible AI Use Cases and Creative Solutions

We often start with this stage and skip or delay the first two parts. Doing the ideation phase as the third step is essential for you to better understand how people feel and what the human problems really are before you start solving. It is important in a design session to ideate with mission, purpose, and values being front and center. The output of this phase is likely to be a list of solution concepts to advance to prototyping.

Prototype: Design Small-Scale Pilots with Checkpoints and Ethical Guardrails

At this stage, ideas turn into a reality with the prototype phase. For AI integration, this means creating a testable and tangible version of the solution. It would be small-scale with the mantra of building to think. In a nonprofit AI context, a prototype could be as simple as a spreadsheet or as complex as a beta test of an AI tool. Checkpoints and guardrails should be built into the prototypes so that you know when to advance to the next step, what you are testing for, what you are collecting, and when to pull the plug on the project. Another key theme here is to fail fast. This does not mean to move fast or haphazardly, but rather the ability to fail fast, learn from that, and move to the next iteration.

Test: Gather the Feedback, Measure Trust, and Refine Your Work

The testing phase is when the prototype comes to life and engages with the reality of day-to-day usage. This means trying out the AI solution or pilot in the real world, but in a controlled environment. This is a chance to test different scenarios and collect feedback from everyone involved. Success in advancement settings is measured in trust, adoption rate, and results. Trust is a key and essential ingredient, and the empathize and test phases bring that to the front of the process. Testing should also include those affected indirectly or any byproducts involved in this process. In this phase, even negative results are extremely valuable, as they inform the next iteration of the process and bring you closer to the solution or help refine the problem.

INSTITUTIONAL DESIGN

Among the many institutions that designed unique models for AI use, below are four examples highlighting stories of creating an AI roadmap (Cornell University), AI principles

(Stanford University), AI structure (MIT), and AI executive coordination (Brown University).

Roadmap for Responsible Design: Cornell University

In early 2024, Cornell University became one of the first major institutions to turn the generative-AI surge into a governing design.[10] While many peers debated policy statements, position statements, or drafted short-term guidelines, Cornell took the practical route to build a structure.

Commissioned by then-Provost Michael Kotlikoff, now president of Cornell, the Generative AI in Administration Task Force (GAATF) produced a 30-page report that remains one of the clearest frameworks for integrating AI into the daily operations of a university. The charge was clear: Evaluate AI use in administration, from admissions to HR, finance to student services. The goal was to create guardrails that encourage experimenting in a safe environment and not red tape that smothers innovation.

The committee represented every major function: legal counsel, IT, data governance, communications, academic affairs, and risk management. The design principle was that innovation requires governance to scale safely.

Designing Governance by Principle: Stanford University[11]

If Cornell University drafted the roadmap, Stanford University wrote the manifesto. In January 2025, many in the nonprofit sector, particularly in higher education, took notice when Stanford released its "AI at Stanford" Advisory Committee report.[12] Here was another major university acknowledging that AI's rapid rise demanded that we take action and be proactive, not in the distant future, but right then.

Formation

The formation of an AI committee in 2024, its mode of operating, and its resulting principles offer a masterclass in moving from disruption to design. It vividly demonstrates how an organization, built on the work of peers like Cornell's AI Task Force, applied change frameworks to create a governance blueprint for the institution. Let us unpack how Stanford designed its AI strategy and what nonprofits can learn from it.

Genesis

In early 2024, the Stanford provost charged a new committee with evaluating the role of AI in the university administration, education, and research. This charge was sparked by both excitement and concern. Faculty and staff were already experimenting with tools in classrooms and back-office tasks, but with no overarching policy or clear guidance. The university's leadership wanted to design an approach before unintended consequences spiraled. The committee was intentionally cross-functional and included 10 members, including faculty members from engineering and medicine, staff from IT and ethics, and representatives from other key units.

Stanford's committee decided to build upon the March 2024 report from Cornell University showing a collaborative ethic: This was not a competition, but a shared quest for responsible use of artificial intelligence in higher education.

From Policy to Principles

Chaired by bioengineering professor Russ Altman, the 10-member committee crafted six guiding principles over four months and seven meetings:

1. Human oversight and accountability
2. Alignment with institutional values

3. Professional responsibility
4. Ethical use and transparency
5. Privacy and security
6. Data quality and reliability

Dr. Altman's most memorable quote was the "AI Golden Rule": *Use or share AI outputs as you would want others to use or share them with you.*

This single sentence distilled the human-centered philosophy of the committee and is consistent with the spirit of Stanford's approach to AI in general. After all, they created the Human-Centered Artificial Intelligence Institute in response to the rise of AI.[13] (While this institute is more on the academic and research side of Stanford, it is indicative of the implementation at the university level.)

Culture as Design

The university launched an AI playground and sponsored forums where students, faculty, and staff debated real scenarios. The members of the committee became visible ambassadors visiting departments and schools to discuss both risk and opportunity. By embedding ethical reflection in daily conversations rather than in compliance and forms, the university moved towards building collective ownership.

Altman emphasized the role of evangelization in adoption. The report itself was brief, only 20 pages, starting with principles and speaking plainly.

By codifying shared language and modeling transparency, organizations can turn ethics from an abstract conversation into habits. Its AI Golden Rule became shorthand for an era of design grounded not in fear of AI, but in faith in human judgement.

Structural Change: MIT's Designing of a Bilingual Future

In 2019, the Massachusetts Institute of Technology (MIT) undertook a major structural redesign that was kind of unique in its time. They announced the creation of the Stephen A. Schwarzman College of Computing (SCC).[14] This was not a gift that resulted in adding a donor name to a school or a college; it was a partnership that included a major philanthropic gift of $350 million from Mr. Schwarzman to create new architecture for learning in the age of intelligence.

The Vision

The vision was articulated by its chief architect, MIT President Rafael Reif: "To educate bilinguals who are experts in computing and in another field of inquiry." Computing would no longer be a discipline; it would be a language in a way, the new math. Every MIT student, including biologists, architects, economists, or designers, would graduate fluent in AI and ethics.

Structural Innovation

The college introduced three design choices:

- Integration across schools:
  - Each department within MIT built a bridge to the new college, and SCC was embedded within MIT's five existing schools.
  - Each department would develop its own "Computing + X" track.
- Joint faculty appointments:
  - Professors held dual roles between their home departments and SCC, allowing knowledge to flow freely across disciplinary lines.

- Ethics in every course:
    - The Social and Ethical Responsibilities of Computing (SERC) initiative required faculty to embed ethical reflection within technical curricula instead of isolating it in a single class.[15]

The design was as symbolic as it was administrative. It signaled to the community that computing belonged everywhere, and that moral literacy was part of technical competence.

Impact

Five years later, the results are visible across MIT. Hundreds of students graduate every year from Computing + X tracks, and joint research grants between multiple disciplines are happening more organically.

For nonprofits and mission-driven organizations, MIT's lesson is scalable: Readiness requires bilingual design. The future team members must be fluent in data and purpose, analytics, and empathy. This effort requires structure, clear pathways, shared incentives, and visible values.

MIT did not adapt to AI; it rearchitected for it. By turning computing into a connective infrastructure, it shows that innovation and governance can coexist when design aligns technology with human intent.

Executive Coordination: Brown University

By mid-2024, Brown University extended the AI design story into a new phase: institutional orchestration.[16] Brown decided to connect it all through leadership. The university appointed computer science professor Michael Littman as its first associate provost for artificial intelligence. Unlike task forces or colleges, this was an executive role inside the cabinet of the provost (Brown's chief academic officer), a permanent senior position charged with coordinating AI strategy across education, research, and administration. In his announcement, Littman remarked: "AI is not one thing happening in one department, it's a transformation that affects how we teach, how we discover, and how we serve society."

Expert Voice: Kerry Evanter

Designing with AI, Not for AI[17]

As the executive director of advancement communications, learning and development at Clark University, Kerry Evanter has a unique perspective on the use of technology in advancement. As she was solving for increasing alumni engagement, Kerry looked less at the messages and numbers and more for the meaning. She has spent the recent part of her career at the intersection of storytelling and data analytics. Kerry is now at Clark University, leading a bold experiment that is attempting to turn alumni engagement into a measurable, gamified ecosystem that rewards authentic participation rather than passive clicks and attendance. What matters most in Kerry's story is not what she developed, despite it being a very innovative idea, but rather the design approach and methodology that she used and that you can replicate.

Mandate to Design

Kerry gave herself the mandate to design as she started overseeing a study with Schelling Point, a consensus design tool named after Nobel Laureate

Thomas Schelling.[18] She was introduced by Clark's head of human resources, Randi Nichols, to the tool that replaces convening endless focus groups with finding natural points of alignment among hundreds of opinions. Through a sequence of surveys, facilitated sessions, and one-on-one interviews, they identified a series of opinions.

During her interview with Clark's president, David Fithian, he challenged her to "find something unconventional that we can do rather than just doing the same things better. This is when Evanter turned to AI by starting to ask ChatGPT about some innovative ideas, and the concept of creating a Points and Tiers system for alumni engagement came to life.

Instead of taking alumni generosity as a given due to loyalty, the proposed program rewards generosity, mentorship, advocacy, and leadership rather than transactions through a robust points system. Alumni can then redeem these points for things that matter the most to them, such as discounts or free micro-credentials, access to mentors, resources, and faculty members, all from within the network. Alumni are rewarded for supporting students by helping with admission leads; providing mentorship, coaching, internships, and job opportunities to students and young alumni; engaging with advancement by hosting or attending events, donating or advocating; and interacting on social media and being ambassadors on behalf of the university. Each action earns points redeemable for experiences, such as meeting at a school, priority registration at events, discounts on alumni experiences or academic offerings, and programs, among others.

Evanter interacted with people to identify problems and sentiments, used AI as an idea generator and sounding board, and then validated the ideas with people to expand the initial AI-generated idea to advance it to a fully developed concept. She provided this concept to the faculty and students of the Clark Becker School of Design and Technology to design it as a product and an app that can gamify that experience. They might or might not use AI tools in the development of the solution, but using AI to design the solution itself is the key takeaway here.

There are a lot of questions that members of the community have to solve, such as how much weight the program puts on each activity, which will reflect the values of the institution and should be ethically guarded.

The program is likely to generate new types of data that will turn into key performance indicators, impacting future dashboards, what we track, and what we will value, like say engagement velocity (how quickly alumni are moving from first contact to sustained involvement), engagement diversity (how many different activities a person chooses), and engagement personas (what type of alumni are using the product and who is left behind).

Deciding whether a pilot is ready to scale or should be declared a failed pilot is another important decision that could be AI-informed. Soon, Evanter will face that crucial decision.

DESIGNING FOR TRUST AND IMPACT

Let's zoom out and consider the ultimate goal of all this design work. At its heart, designing for responsible innovation is about building trust, whether with your supporters, within your team, or with the public. If you take time to methodically design how AI fits into your mission, you are telling your stakeholders: We are being careful stewards of this powerful technology. Just as Stanford University did by publishing principles and inviting a conversation around them, Cornell by creating sandboxes and a transparent review process, Brown by appointing a dedicated leader, and MIT by redesigning its whole org chart. Plainly put, trust is the currency that converts technological potential into real-world impact.

Designing for trust also means being transparent about failures, not just successes. Share your views, findings, blueprints, and guidelines with others in the sector and contribute to the conversation, irrespective of whether you are excited or concerned about this technology. This kind of openness is part of ethical design.

Moving from disruption to design is both a strategic choice and a moral one. It is a choice to steer the ship and not drift with the currents of AI. We have to acknowledge what the MIT report highlighted: AI fails in the organization when people, culture, and processes are undervalued, not in the labs, where the environment is controlled.[19]

Immediate Actions for Leaders

1. Host a design workshop and explore a problem like donor retention or a data entry backlog, and run a two-day design sprint with a core team to generate AI-related solutions.

2. Gather a list of needs and concerns, addressing privacy, bias, and transparency up front.
3. Include members from different teams and give them four to six weeks to prototype and test solutions using design-oriented thinking.
4. Pilot, evaluate, and scale whenever possible.

The small wins and lessons learned in each step will fuel the momentum for your work and bring you, with every pilot, closer to becoming an AI-ready and human-centered organization.

Chapter Four Recap

Core Message

Focus on designing AI systems instead of adding more ad-hoc pilots. Identify your readiness, define what success looks like, and recognize the possible risks you may come across.

Key Takeaways

- Diagnose before you prescribe. Use the AI Readiness State (AIRS) to locate your team's starting point and pick right-sized pilots.
- Design pilots as blueprints. Define success and "kill switches" in advance. Test, evaluate, and document so that you and others can learn from your findings.
- Build governance and ethics into your pilot design. It might take a bit more time, but it will help you in the long run to increase your ethics maturity as you increase your AI readiness.

- Integrate data where it matters and let the Four Data Frontiers framework guide what signals you want to surface.

Why It Matters

By focusing on thoughtful design, your AI future will be immersed in higher adoption, true problem-solving, and easier-to-scale plans.

Chapter Endnotes

1. "Giving USA 2025: U.S. Charitable Giving Grew to $592.50 Billion in 2024, Lifted by Stock Market Gains," Giving USA, June 24, 2025, https://givingusa.org/giving-usa-2025-u-s-charitable-giving-grew-to-592-50-billion-in-2024-lifted-by-stock-market-gains/.
2. Joseph L. Bower and Clayton M. Christensen, "Disruptive Technologies: Catching the Wave," *Harvard Business Review* 73, no. 1 (1995): 43–53, https://www.hbs.edu/faculty/Pages/item.aspx?num=6841.
3. "Disruptive Innovation Theory," Christensen Institute, accessed March 1, 2026, https://www.christenseninstitute.org/theory/disruptive-innovation/.
4. Aditya Challapally, Chris Pease, Ramesh Raskar, and Pradyumna Chari, *The GenAI Divide: State of AI in Business 2025* (MIT NANDA, 2025), https://mlq.ai/media/quarterly_decks/v0.1_State_of_AI_in_Business_2025_Report.pdf.

5. Christopher Cannon (Generosity Collaborative), interview by Joe Manok and David Woodruff, 2025.
6. Greg Eow, Shawn Ferullo, Kendra Leith, Joe Manok, and Lilen Uchima, "MIT Leader to Leader Program," *Change at MIT Study* (Massachusetts Institute of Technology, 2018), https://hr.mit.edu/l2l/projects.
7. Dedric Carter, "3 Lenses for Management Success," MIT Sloan School, March 10, 2014, https://mitsloan.mit.edu/emba/3-lenses-management-success.
8. Stanford University, "A Design Thinking Process," ME113, accessed March 10, 2026, https://web.stanford.edu/class/me113/d_thinking.html.
9. Scott Doorley, Sarah Holcomb, Perry Klebahn, Kathryn Segovia, and Jeremy Utley, "Design Thinking Bootleg," Stanford d.school, accessed March 1, 2026, https://dschool.stanford.edu/tools/design-thinking-bootleg.
10. "Generative AI in Administration Task Force Report (January 2024)," Cornell University, accessed March 1, 2026, https://it.cornell.edu/ai-strategy/aiinadministration.
11. Russ Altman (professor, Stanford University), interview by Joe Manok and David Woodruff, 2025.
12. "Report of the AI at Stanford Advisory Committee," Stanford University Office of the Provost, January 9, 2025, https://provost.stanford.edu/sites/g/files/sbiybj29391/files/media/file/ai-at-stanford-report.pdf.
13. "About," Stanford Institute for Human-Centered Artificial Intelligence, accessed March 1, 2026, https://hai.stanford.edu/about.

14. L. Rafael Reif, "Letter to the MIT Community Regarding the MIT Stephen A. Schwarzman College of Computing," *MIT News*, October 15, 2018, https://news.mit.edu/2018/letter-mit-community-regarding-mit-stephen-schwarzman-college-computing.
15. "Social and Ethical Responsibilities of Computing," MIT Schwarzman College of Computing, accessed March 1, 2026, https://computing.mit.edu/cross-cutting/social-and-ethical-responsibilities-of-computing/.
16. Francis J. Doyle III, "Michael Littman Appointed as Brown's Inaugural Associate Provost for Artificial Intelligence," Office of the Provost, Brown University, December 4, 2024, https://provost.brown.edu/communications/michael-littman-appointed-browns-inaugural-associate-provost-artificial-intelligence.
17. Kerry Evanter (executive director of advancement communications and strategic information management, Clark University), interview by Joe Manok and David Woodruff, 2025.
18. "About SchellingPoint," SchellingPoint, accessed March 1, 2026, https://schellingpoint.com/about-us/.
19. Aditya Challapally, Chris Pease, Ramesh Raskar, and Pradyumna Chari, *The GenAI Divide: State of AI in Business 2025* (MIT NANDA, 2025), https://mlq.ai/media/quarterly_decks/v0.1_State_of_AI_in_Business_2025_Report.pdf.

CHAPTER FIVE

Building the AI-Ready Team from Design to People

The core truth about AI in nonprofits is that we need AI models to predict generosity, and we need people to inspire it. After all, "people give to people," and we cannot lose sight of this guiding principle.

Success is not having the flashiest algorithm or the largest data warehouse: It's about the people behind the technology and how they will use it. In this chapter, we will pivot attention from fundraising systems to the staff using the products and to design for people. We bridge the high-level ethics and governance frameworks of chapter three with the design mindset of chapter four, focusing on the human architecture required to truly become an AI-ready nonprofit.

TALENT OVER TECHNOLOGY: COORDINATION AND THE HUMAN FACTOR

Steven Suda, a veteran principal gifts fundraiser who recently retired from Stanford University, is known for his programmatic wisdom.[1] While the whole world switched to Zoom and became technology-heavy amid the 2020 global pandemic, Suda kept doing what he did best: spending time with his prospective donors, even playing golf with those living in the Bay Area. It just so happened that his way of engaging them was already COVID-proof with fresh air and social distancing. While other fundraisers struggled to adapt, Suda was able to deepen his relationships without using a screen or a single technology. Suda is proof that technology is not always the answer, but designing a better experience for our donors and community is more than often improved with technology.

If you think of fundraising campaigns as an orchestra, AI is a brilliant new instrument, unsure when to join the song without a conductor and a great team. An instrument alone cannot engage and inspire people. By working in tandem with existing processes, AI can provide answers in an efficient way.

Simply put, AI readiness is not just an IT project or adding yet another new vendor; it's truly a transformation of "change management." We must be intentional about AI and maintain, if not supplant, the human connection.

The AI-Enabled Professional: Seven Essential Competencies

Closing the talent gap begins by defining exactly what a successful AI-Ready professional looks like. What new skills and mindsets do staff need to thrive alongside intelligence machines? The Business Higher Education Forum developed a set of seven essential skills identified as critical in the age of AI in 2025:[2]

- **AI Literacy:** A baseline understanding of key AI concepts and tools. They don't have to be scientists, but they should be able to grasp how AI works, its limitations, and its capabilities. They must understand what an algorithm output means and what it might not mean, rather than seeing it as a "black box."

- **Data Literacy:** The ability to understand and work with data is equally important. Data is only beneficial if it can be analyzed and applied to make informed decisions. Are the staff trained to interpret the insights and identify data quality issues?
- **Critical Thinking and Creativity:** As AI systems are rapidly improving, we still need the human element in fundraising. Staff should understand where machines fall short and apply human creativity.
- **Ethics and Responsible AI Use:** Nonprofits should establish a solid grounding in data ethics, privacy, and their organization's particular AI governance and policies, including concepts like bias, consent, and transparency.
- **Digital and Computational Skills:** It can be difficult to keep up with advancements, but if it's not enough to adopt modern tools such as automation, software, and basic coding, staff should be able to navigate the automated, digitized systems.
- **Collaboration and AI Era Communications:** Building on their interpersonal skills, staff are faced with a new challenge: collaborating with AI or using AI in a seamless way.
- **Adaptability and Continuous Learning:** Given the pace at which change is happening, specific tools will come and go, making it crucial to learn how AI works and how to embrace this transformational change. Staff should be encouraged to adopt a lifelong learning perspective.

These seven competencies form the scaffolding for hiring and training in an AI-driven nonprofit. They reflect a blending of hard and soft skills, a sort of technical literacy and humanistic insight.

Core Capabilities: Rethinking Organizational Structure

While nonprofits and businesses are structured around departments, fundraising, marketing, IT, etc., AI must be infused throughout the organization. Explore the following five core capabilities, and have teams align around them.

- **Insight:** Develop the ability to generate data-driven foresight and insights. This encompasses prospect research, analytics, market intelligence, and other ways we learn from information. From an AI lens, this would include predictive modeling, natural language processing, or even enhanced reporting. Some organizations are starting to embed analysts in their teams and working on their "upskilling" strategy.
- **Activation:** Commit to deploying these insights in day-to-day work.
- **Governance:** Establish the processes and structures that oversee the use of AI to ensure alignment with institutional values and ethical principles.
- **Enablement:** Ensure that the infrastructure and guidelines can position the team to be AI-enabled, including training programs and support services.
- **Measurement:** Evaluate the metrics to determine impact and learn from what the data tells us. This involves clearly defining what success looks like and defining ways to monitor and evaluate it.

Instead of thinking about AI as a separate tech project, imagine it as strengthening these core muscles of the organization. Equally important, these capabilities cut across the traditional organizational chart and help break silos and bring stakeholders together.

Organizing Models for AI Readiness: Hub and Spoke, Community of Practice, and Innovation Labs

With a strong AI foundation built from the previous chapters, the question becomes: How should an organization be structured to maximize the chances that an AI project succeeds and AI skills propagate? There is no one-size-fits-all, so reflection and self-evaluation are critical here. What's the culture? How would the team react? There are several models that have gained traction:

Model 1: Hub and Spoke (Center of Excellence)

In this model, a central team concentrates AI expertise, while the spokes are the various units or departments that the hub supports. The hub typically sets the standards, provides guidance, develops core tools or models, and sometimes handles the most complex tasks (like building a custom AI model). This is particularly helpful if skilled personnel are limited and you want to centralize them in one place. This could be as small as two to five people or more, depending on the size of your organization. These people serve as the hub and are helpful in either vetting or contributing to your internal vetting processes of third-party AI tools. The spokes are the units within your advancement team, or for smaller organizations, other units in your nonprofit or on campus.

Model 2: Community of Practice

This networked approach relies on voluntary communities of practice rather than a strict hierarchy. The idea is to create forums where people interested in AI can share their knowledge, collaborate on ideas, and spread best practices across silos. These could be other units within your advancement team, others across your nonprofit or campus, or even others in the industry who have similar roles at other institutions.

Model 3: Innovation Lab

The third model consists of creating an AI lab, often with a corporate partner, and uses that as the dedicated virtual, physical, or organizational space for experimentation and prototyping. Clark University Advancement has adopted this model and works with EverTrue. Compare it to a mini startup within your organization focused on reviewing and deploying emerging technology, like AI, for the mission. The labs are often tasked with piloting new ideas in a low-risk environment, then spinning off successful projects into the main organizational structure. The lab itself might be funded through philanthropy, as some donors or partners enjoy innovation and forward-thinking nonprofits. The risk of labs is the disconnect from the core operations; the "cool stuff," meaning the tools and emerging ideas, may remain isolated and become harder to integrate into business processes quickly, so it is important to include the end users (like the fundraisers and researchers) in the design process and the lab itself. The key advantage is that an innovation lab creates a close environment to ideate, pilot, test, fail fast, and pivot, then scale when things work out.

Whatever model you decide to use, it is important to spend time with your team and leadership to assess which model would best suit your organization, your culture, and your needs. Be thoughtful and actively select a model as you start your AI work or scale it.

JD 2.0: Job Descriptions for the AI Era

The rise of AI has necessitated AI capabilities in job descriptions. Fast Company recently noted a trend toward "AI translators," who can help decision-makers better understand AI insights and ensure they are applied correctly.[3] This requires reimagining job descriptions and rewriting them to explicitly include AI-related responsibilities and skills. For example:

- A gift officer's job description might include leveraging AI-driven prospect intelligence to prioritize outreach.
- A data/researcher role might evolve to include "develops and maintains predictive models for donor behavior" or "ability to generate, evaluate, fact check, and edit AI-generated research profiles."
- A marketing director might add: "utilizes AI content generation to optimize messaging or ensure brand consistency."
- New titles could appear as well, such as an AI implementation specialist or ethical data officer.

Recruitment strategies must adapt to identify AI experience in conjunction with industry-specific skills.

The models are a good introduction, but in order to understand what is needed to build an AI-Ready team, let's explore living examples and profiles of professionals and organizations:

CUSTOM GPTS, VOICE SAFETY, AND CREATIVE COLLABORATION

Ashley Budd, Cornell University

"If you don't have authenticity, empathy, and logic in your messaging, people won't trust it," says Ashley Budd, senior director of advancement marketing at Cornell University.[4] In her role, Budd is responsible for engaging alumni and donors through digital campaigns, email, and social media.

The author of *Mailed It*, Budd brings a wealth of AI experience to her job every day, quietly transforming her team's approach by embracing AI as a creative collaborator. Her team manages hundreds of campaigns a year, from appeal letters to event invitations. Each carries Cornell's trusted voice, warmth, and transparency with donors.

Budd soon encountered a common challenge: The volume of communications skyrocketed, bringing a greater need for more personalized emails, more donor reports, and more social media posts. But her team size was not growing. Instead of using traditional ChatGPT prompts, Budd decided to tell her experience and create a custom GPT. As the name implies, these are more specialized, generative AI tools that focus on a specific set of advanced instructions for a defined set of projects, assuming a set of responsibilities for the identified audiences.[5] This resulted in designing a Budd-onboarded GPT trained on Cornell-specific content for the first draft. She then reviews the work and refines it. Budd refers to the human review process as voice safety. Just like brand safety and advertising, voice safety with AI means ensuring the institutional tone, facts, and values are correctly represented by AI. Aside from the increased output, alumni reported feeling that communications were more tailored to them. Budd attributes this to freeing her staff time to focus on the content, strategy, and personalization details rather than spending much of their time on basic drafting and copywriting.

Kim Infanti, Syracuse University[6]

"I'm very happily married, but I have a 'boyfriend' named Claude." That's how Kim Infanti, executive director of alumni constituent engagement at Syracuse University, opened the 2025 AI talk. Her humor perfectly captured the world we're living in: One where we spend a lot of time chatting with AI agents.

Claude is Anthropic's AI assistant that makes Infanti's workday more efficient, and that is just the beginning.[7] In 2024, Syracuse launched campus-wide access to generative models for education and operations. Within weeks, Infanti saw potential for beyond classroom use and basic operations optimization. It seemed like the solution to the challenges her communications team faced (the same as Cornell's): more stories, more events, fewer people.

Rather than view AI as competition, Infanti introduced it as a creative collaborator, a sort of colleague. She began documenting every practical use case, eventually compiling a top 10 list that she now teaches to her peers. They range from video scripting and event planning to survey analysis and social content creation. By providing an outline, Claude is able to generate a first draft, which Infanti and her team rewrite, an approach that relieves the heavy lifting while keeping the human touch.

Perhaps Infanti's boldest experiment so far was with the Syracuse 2019 survey archive of more than 300 pages of open-ended responses. She uploaded them to Claude, and in 22 seconds, it produced the responses into a clear summary with action items. The results informed new messaging and programming priorities for 2026.

Equally transformative was Infanti's adoption of Protopia's Ask Orange Alumni, an AI-driven platform that matches student questions with alumni experts.[8] The "human connection facilitated by machine intuition," as she puts it. Since its launch on May 20, 2025, thousands of micro relationships have formed through automated matching, each one a small bridge between generations of the Orange Network.

Although AI can be overwhelming, Infanti has made it work for her team. What began as a novelty has become a professional development pipeline. Within one year, Syracuse's advancement division increased content output by 50 percent with no additional staff. Response rates on segmented campaigns rose by 18 percent, and internal surveys show a 40 percent decline in reported creative burnout. But perhaps Infanti's real victory is cultural. She concluded: "People are not afraid of AI anymore. They are curious, and that curiosity keeps us improving." Syracuse's story demonstrates that when a leader models humanity, transparency, and a little humor, technology can become a teammate and not a threat.

OPERATIONS AND DATA: DATA ETHICS AND RESEARCH INTEGRATION

Amy Gingle, Orr Group[9]

Amy Gingle offers a special perspective as a senior director at Orr Group. Her superpower lies in prospect research and analytics from her work at Massachusetts General Hospital and Year Up, large organizations where she built prospect development functions.

In her consulting role, Gingle found that many nonprofit teams are either intimidated by AI, "It will ruin donor relationships or take their jobs," or are unrealistic about it, "AI is going to fix fundraising problems overnight." Gingle emphasizes a strong prospect management process first, which includes clear criteria for prospect prioritization, cultivation plans, and tracking, essentially having order and avoiding chaos. She then introduces AI or advanced analytics to enhance those processes, melding old-school fundraising with innovation. Her guiding principle is that technology should free people to spend less time on burdensome administrative tasks and unnecessary searching, and more time on strategy and building relationships.

Gingle describes prospect research as "the OG of AI for fundraising," noting that fundraisers and researchers have long applied similar principles of pattern recognition, inference, and segmentation that now underpin modern machine learning. In her recent work, she demonstrates how new tools from DonorAtlas to ChatGPT can rapidly generate transparent, cited prospect insights while still relying on professional judgment and ethical review.

Orr Group's pilot systems integrate AI and automation to create draft donor profiles through linked platforms such as Slack, Salesforce, Foundation Directory, Appolo.ai, iWave, and Cause IQ. AI pulls the relevant data and automates it into a standardized profile. Researchers then review, verify, and refine the output to ensure that every profile remains human-informed, accurate, and compliant with approved data sources. Finally, automation uploads the completed profile into CRM records, keeping the database timely and consistent, and prospect data easily accessible.

Gingle urges advancement teams not to get caught up in the hype. Instead, she encourages them to pilot thoughtfully, understand their comfort level with new tools, and choose what aligns with their culture and goals. As Gingle often reminds teams, there's no single "right" way. Her clear message embodies our hope for this book: Effective AI adoption depends on transparency, ethics, and human experience. In Gingle's view, AI should be applied with intention, and its real value emerges when it enhances the work rather than serving as technology for technology's sake.

FUNDRAISING AND COACHING: BEHAVIORAL COACHING AND FRONTLINE ENABLEMENT

Mallory Erickson, Practivated[10]

Frontline-facing colleagues from gift officers to donor relations staff are often where the rubber meets the road for any AI innovation. If a new tool or approach does not resonate with them or isn't understood, it simply won't get used, and all the data science in the world won't raise an extra dollar. That is why the intersection of fundraising and coaching is so critical. Here is where Mallory Erickson excels as a fundraising leadership coach and innovator, examining the behavioral and psychological side of fundraising. She helped found Fundraising.ai, is the host of *What in the Fundraising* podcast, and founded Practivated, an interactive AI training program to help fundraisers overcome common challenges.

Erickson emphasizes that most fundraising problems are not due to lack of technical knowledge, but rather issues related to anxiety, lack of confidence, poor habits, inconsistent execution, disorganization, and burnout. After all, practice is key, preferably not "live" in actual donor meetings, which can be stressful, suboptimal, and very costly for organizations. Erickson tried to convince several companies to create a practice product, and when that was unsuccessful, she created a company to bring the idea to life: Practivated, an AI-powered donor conversation simulator. The tool plays the role of a coach; fundraisers can log in and engage in a simulated donor dialogue, powered by AI, that responds as a real donor would. This allows them to try out different ways to make an ask or tell a story using real-life scenarios. The system then provides feedback on elements like tone, clarity, emotional cues, and so forth.

Think of it as a flight simulator. The AI models do not use any donor-specific information, but rather natural language processing to analyze the fundraiser's spoken or typed responses, and can be tuned to simulate different donor personas (for example, the hesitant or skeptical analytical engineer, the chatty alum who goes off on tangents, the inspiring leader, and so forth).

Erickson, interested in the psychology of asking, authored the book *The Courage to Ask*. In a recent research project performed at Practivated, she and her team analyzed 370 fundraising conversations and revealed the behavioral and conversational gaps that make fundraising feel harder than it should. The data showed that while the fundraisers were warm, authentic, and deeply mission-driven, they made clear, confident asks less than half of the time. And "asks" doesn't just mean solicitations for philanthropic support; this includes any further steps, such as scheduling a follow-up, requesting a decision timeline, or proposing a visit.

Her team found that two critical breakdowns drive this pattern. First, the ask itself often included softening or apologizing throughout the moment. Second, the flow of conversation was at 54 percent effectiveness, showing many struggles with transitions of when to shift from rapport to purpose, or how to navigate awkward pauses without losing momentum. The result is what she calls the "friend zone of fundraising," where conversations feel good but lead nowhere. Her insights reframe fundraising as a behavioral skill and not a messaging script. Fundraisers, according to Erickson, do not need more words, but tools to manage hesitation, read emotional cues, and move conversations forward with confidence and integrity. Practivated's approach to coaching builds these micro-skills through practice and guided reflection, showing leaders how to identify where their teams get stuck and when to intervene without micromanaging.

By translating behavioral data into actionable coaching frameworks, Erickson is redefining what it means to be an AI-era fundraiser as someone who blends emotional intelligence, data-informed insights, and human courage to ask for what really matters.

Matthew Letendre, Dana-Farber Cancer Institute[11]

Matthew Letendre is senior director for principal gifts at Dana-Farber Cancer Institute (DFCI). He represents a new class of fundraisers who are adapting to technological advances by using artificial intelligence not as a theoretical opportunity, but as a practical tool for better preparation, sharper communication, and deeper personal connection.

A veteran of EverTrue, Villanova University, Boston University, and Brown University, Letendre joined DFCI to help one of the world's leading medical research, patient care, and fundraising organizations reach its full potential. He views the ethical use of AI as a key contributor to that effort.

At the time of writing, Letendre was exploring AI in practice and using it in his daily life to augment his capabilities, not to replace himself or his colleagues. When he arrived at DFCI, he inherited (as fundraisers often do) a large and complex portfolio. His challenge was not a lack of information, but a lack of time, and the many hours spent on manual tasks: summarizing profiles, drafting outreach messages, logging contact reports, and preparing meeting agendas. Letendre's first step was to look inward, asking how much of that work could be supported or accelerated through AI and automation without sacrificing authenticity.

DFCI had already invested in a generative AI environment that is a secure, HIPAA-compliant, and auditable platform built for research, operations, and administrative applications. The system runs within DFCI's protected network and mirrors the data governance standards used for clinical research. This environment became Letendre's foundation. He helped his division adopt the platform for fundraising use, ensuring that the private and auditable AI environment served the philanthropy team in their job functions.

Letendre introduced these capabilities under a simple framework: "Meet my team," he said, introducing four AI agents, each with a specific role-based function within the gift cycle:

- **Research Agent:** A web-enabled assistant specialized in producing concise, citation-based donor research profiles. It can pull structured data from internal systems and public sources, then summarize them into a one-page format.
- **Outreach Agent:** A writing partner trained to draft personalized donor emails and event invitations, using CRM tags and recent interactions to keep tone and timing appropriate.
- **Prep Agent:** A meeting prep assistant who condenses research and relationship history into talking points, key priorities, and open questions for upcoming donor visits.
- **Report Agent:** A CRM integration helper that turns meeting notes into compliant contact reports, capturing action items and next steps in the correct institutional language.

This AI team did not replace staff; it extended them. The team effectively gained a small suite of virtual colleagues that handled the mechanical work, allowing him to focus on human conversations, driving philanthropic relationships. Letendre's framework also modeled responsible implementation at every step, adhering to DFCI's data governance policies and excluding any patient confidential information. Each prompt and output was auditable, and the AI system logged all user sessions, allowing compliance officers to review interactions if needed.

This transparency reassured leadership that innovation would not come at the expense of institutional integrity. Within months, Letendre reported tangible improvements: faster administrative turnaround times for profiles, increased engagement from a relatively unknown portfolio, and dramatically shortened contact reports. He entered meetings better prepared and left them more confident in the next steps.

Letendre is a strong believer in the practical use of AI and encourages fundraisers to experiment with the technology, where it can improve efficiency without sacrificing personalization and authenticity.

Erickson's and Letendre's approaches embody AI enablement at its best. The result of their AI use and automation is not depersonalization, but greater personalization. Fundraisers became more attuned with donor cues and more confident in authentically responding because they had drilled those essential skills with an AI partner.

STEWARDSHIP AND INNOVATION: FROM STAFF DESIGN TO COMPANY LAUNCH

Chelsea Lamego, FundMiner[12]

Chelsea Lamego exemplifies how identifying a gap in organizational capability (in her case, around donor stewardship and fund management) can spark not just internal innovation, but an entirely new venture that benefits many organizations, including her own.

Lamego's journey from the assistant vice president for advancement operations at the University of Texas at El Paso (UTEP) to co-founder of FundMiner highlights the power of applying design thinking to people and processes. This underscores that innovation in the nonprofit sector often originates from those who deeply understand the work, the pain points, and the opportunities.

At UTEP, Lamego was responsible for the actions taken after the gift was received, ensuring that the funds were used in accordance with the donors' intent, maintained compliance with reporting, and demonstrated impact back to the donors (stewardship of the relationship). This is an unsung but crucial component of fundraising and the main reason why people give repeatedly or never give again to an organization. Lamego noticed a systemic issue: Across higher education institutions and nonprofits in general, fund management was often reactive, manual, and siloed. She saw millions of unused dollars sitting in funds because program staff didn't know the money was available or didn't spend it in time. She saw donor reports going out late or being generic, risking trust. She and her team also drowned in spreadsheets, attempting to match donations to expenses, impact reports, and stories. Most organizations say their fund management is ad hoc or reactive. UTEP was not alone, which Lamego later confirmed by a survey that she conducted.

As a forward thinker, Lamego applied design principles by empathizing with stakeholders and defining the problem as a lack of an integrated, user-friendly system for fund stewardship. She then ideated solutions for a central platform that pulled financial data, tracked compliance, and even nudged staff when actions were needed. Next, she prototyped an in-house gift and endowment management system as a central platform. It was modest at first, built on top of their database with custom queries and workflows. For the first time, they had a dashboard of all funds: those with unspent balances, reports due, and so forth.

They standardized reporting templates and started including more engaging data on fund impact. The system became so good that in 2020, it won a Grand Gold Award for Innovation from the Council of Advancement and Support of Education. By creating this tool, Lamego also shifted people's roles, with stewardship officers spending less time as data collectors and more time as strategic partners showing impact stories. It was a new operating model around insight, activation, enablement, governance, and measurement, touching all five capabilities.

Having proven value to UTEP, and with help from the university, Lamego went on to launch FundMiner, a productized version of the concept with AI enhancement. The company uses AI to predict which funds are at risk of under-utilization and alert staff, to match expenditures to the right fund sources, or even to draft elements of donor impact reports by pulling relevant data and narratives. By 2023, FundMiner had gone through the Techstars accelerator, indicating strong market validation.

Lamego's journey from design to product underscores one more lesson: The AI-ready team continuously evolves and sometimes spawns entirely new roles and enterprises. Some staff will become "innovation champions" whose skills go so far that they start training others (within or beyond their organization).

Learning to Embrace the "40th Attempt"[13]

You are likely familiar with a certain blue and yellow can that is found in most homes: WD-40. The acronym is so commonplace that we rarely consider the full name: Water Displacement, 40th formula. This household item, originally created as a corrosion-prevention solution for missile parts, failed 39 times before the formula was successful; this was then honored in the product name. Their market for rocket maintenance was so small compared to the vast world of squeaky doors, rusty hinges, and everyday fixes.

So why does this matter for building an AI-ready team? Because the culture needs to embrace the idea of failed iterations. Your team might need 39 tries to get to the 40th breakthrough. A "failure budget" is key to encouraging experimentation and treating each iteration as a learning step, not a setback. Sometimes the real value is not solving the original problem or use case, but in how a team learns to pivot or repurpose what they've built. In other words, let's not just build teams that "get it right" the first time.

PEOPLE AND CULTURE: PSYCHOLOGICAL SAFETY AND TRUST AS INFRASTRUCTURE

All the cutting-edge algorithms in the world won't transform an organization that isn't culturally ready to use them. Readiness starts with an invisible foundation: psychological safety and trust. By enabling staff to take risks, learn openly, and adapt, you create a solid infrastructure for your AI-ready team.

Research confirms that teams excel when members feel safe to speak up and experiment without feeling embarrassed. Google's landmark Project Aristotle study, which reviewed 180 teams composed of engineering and sales staff, identified psychological safety as the number one factor in high-performing teams.[14] This enables a climate of trust where staff are far more likely to share honest insights, catch problems early, and propose innovative solutions. In short, psychological safety is therefore not "nice to have" but a necessary foundation for everything that matters: innovation, quality, resilience, and transformation.[15]

Forward-thinking companies like Google and Amazon have long embraced a fail-fast ethos, giving employees freedom to test ideas and learn from both successes and mistakes. Nonprofit teams can adopt a similar lab culture mindset. One AI governance study emphasizes that employees should be "encouraged to experiment with AI and learn from both successes and failures," with leadership giving them the freedom to pilot new AI solutions and continuously improve upon them.

Cultivating this culture of experimentation, backed by psychological safety, makes it more likely that your organization will find new and mission-aligned uses for AI and avoid playing it too safe, a problem that resulted in our industry being so behind the curve in reinventing itself and changing how we serve our communities and donors.

Overcoming Fear and Resistance to Change

Even with a supportive culture, leaders must actively address the very human fears that can slow AI adoption. As with most technology, advancement professionals often harbor private anxieties about AI. Some frontline fundraisers, researchers, associates, and communications professionals worry that automated outreach, AI-enabled research tools, and virtual engagement officers are going to make their roles obsolete. (These concerns are common across sectors, not just ours.) In a 2025 survey of CEOs, HR

Dive found that 45 percent of CEOs observed that their employees were being resistant or even openly hostile toward AI initiatives.[16]

A host of barriers underlie this resistance. Fear of obsolescence (Will AI take my job?), lack of trust in AI systems (Will the algorithm make biased decisions?), or simply change fatigue, being comfortable and used to the status quo. Any nonprofit leader thinking of advancing an AI agenda should acknowledge these anxieties upfront.

The fear of job loss must be met, not with dismissal, but with upskilling the role AI is playing and expected to further play in our sector. A culture of learning and reskilling not only helps us face the AI wave but also makes us future-ready for the next challenge, becoming, in a way, an anti-fragile team.

Lack of trust in AI can be met by prioritizing transparent and explainable AI tools and involving end users in vetting them. For instance, a gift officer can be part of vetting a predictive analytics system and be part of the training that explains how the system works. This would make it much easier for them to better understand how the system works and how it makes its recommendations. ChatGPT managed to increase its popularity significantly when it added a transparent process; this allowed users to get an answer *and* the thought process of the algorithm, as well as the webpages and sources for the content.

Finally, effective change management in this context also requires clear, honest communication and strategic planning. Bringing the team early in the AI journey would go a long way as they would be learning with you, evolving their thinking with you, and adapting to change as you are building your plans, as opposed to being brought in a separate deployment or implementation phase.

12-WEEK UPSKILLING SPRINT: FROM DESIGN TO PRACTICE

Imagine condensing a graduate course in AI for nonprofits into three months of hands-on learning experience for your staff, all while they continue to do their work in their existing roles. That is the idea behind a 12-week upskilling sprint.

Many organizations cannot afford to send their entire team back to school, even though many of us work at one, but at the same time, they can't afford *not* to invest in rapid skill-building. A focused quarter-year program can galvanize a transformation from within, creating early wins and internal champions.

At the time of writing this book, Clark University Advancement was initiating this upskilling sprint, so we get to examine it as it rolls out. Keep your eyes open for the outcome and data on this course at Clark. In a 2024 initiative by EDUCAUSE named "AI for Workforce," they noted that "no institution can afford all the staff they need. AI has the potential to augment and assist staff."[17]

So, what does a 12-week upskilling program look like? Here is a sketch that any nonprofit can tailor to their needs:

Weeks 1 and 2: Foundations and Framing

Kick off with training on AI basics and your organization's AI vision. This could involve a series of interactive workshops covering ethical considerations and case studies from other institutions. An executive sponsor should articulate why the institution is adopting AI, address fears, and build excitement around it.

Weeks 3, 4, and 5: Skills Pathways by Role

Using internal talent (if available), external trainers, or online platforms like LinkedIn Learning, assign staff to role-specific learning paths. For instance, frontline fundraisers can focus their path on AI in relationship management, marketing/communications staff can focus on AI for content creation and audience segmentation, and so forth. These pathways can relate to existing courses. The idea here is not to turn every team member into a computer scientist, but rather to give each person a concrete sense of how AI can augment their work.

Weeks 6, 7, and 8: The Hands-On Sprints

Shift gears here from learning to doing hands-on projects, preferably fundraising-related and cross-functional.

Weeks 9 and 10: Synthesis and Upskilling Sprint "Showcase"

Share results from the project sprints and consolidate lessons. Conduct a demo day, almost like a science fair, and bring outside audiences, like institutional leadership, representatives from other nonprofits, or corporate partners. This can be a good opportunity to recognize standout contributions and potentially identify any "AI Champions" in different departments.

Weeks 11 and 12: Integration into Workplans, Job Descriptions, and Next Steps

The final phase ensures this is not a one-off burst and establishes how each participant can create a personal development plan or work plan to integrate the new skills or follow-up actions. On the institutional level, this is a good time to make a permanent support group (perhaps an AI working group) to help promote a culture of consistent upskilling within your organization.

In the next chapter, we will zoom into campaigns and revenue in the age of intelligence while seeing how the AI-ready team can take action, driving innovative fundraising campaigns and revenue models.

Chapter Five Recap

Core Message

Transforming our sector using AI is a transformation project and not a technological one. It requires focusing on the human team rather than the AI tools and vendors.

Key Takeaways

- Tools can amplify human creativity only when supported by the team. Prioritize the seven essential AI competencies.
- Recognize the capabilities over departments: The five organizational capabilities (insight, activation, governance, enablement, and measurement) should replace rigid departmental silos and form the muscles of an intelligent advancement office.
- Professionals succeeding in the nonprofit sector are deploying AI to increase personalization, stewardship, and insight without losing humanity.
- AI resistance should be met with empathy and upskilling, and not counter-resistance.
- Learn from every pilot and attempt, document your failures and successes, and learn from both.
- Plan a 12-week upskilling program so your team is getting more than just a conference and less than a full degree.

In Practice

Investing in both people and platforms is how you reach better adoption, more trust, and how you reach an AI-ready team.

Chapter Endnotes

1. Steve Suda (former principal giving officer at Stanford University), interview by Joe Manok and David Woodruff, 2020.
2. "Future-Proofing Talent in an AI-Powered World: A Sharp Competitive Advantage," Business-Higher Education Forum, September 24, 2025, https://www.linkedin.com/pulse/future-proofing-talent-ai-powered-world-5odqc/.
3. Sarah Bregel, "Postings for This AI Job Are up 800%," *Fast Company*, November 5, 2025, https://www.fastcompany.com/91435680/postings-for-this-ai-job-are-up-800.
4. Ashley Budd (senior director of advancement marketing, Cornell University), interview by Joe Manok and David Woodruff, 2025.
5. "Introducing GPTs," OpenAI, November 6, 2023, https://openai.com/index/introducing-gpts/.
6. Kim Infanti (executive director of alumni constituent engagement, Syracuse University), interview by Joe Manok and David Woodruff, 2025.
7. "Introducing Claude," Anthropic, March 14, 2023, https://www.anthropic.com/news/introducing-claude.
8. Justin Schaefer, "Enhancing Campus Collaboration Through AI-Powered Alumni Engagement at Syracuse," accessed March 2, 2026, https://protopia.co/webinar-case-study-syracuse-university/.

9. Amy Gingle (senior director, Orr Group), interview by Joe Manok and David Woodruff, 2025.
10. Mallory Erickson (founder, Practivated), interview by Joe Manok and David Woodruff, 2025.
11. Matthew Letendre (senior director for principal gifts, Dana Farber Cancer Institute), interview by Joe Manok and David Woodruff, 2025.
12. Chelsea Lamego (founder, FundMiner), interview by Joe Manok and David Woodruff, 2025.
13. "Fascinating Facts You Never Learned in School," WD-40, accessed March 2, 2026, https://www.wd40.com/history/.
14. "Understand Team Effectiveness," Google re:Work, accessed March 2, 2026, https://rework.withgoogle.com/intl/en/guides/understanding-team-effectiveness/.
15. Michelle Bonterre, "Why Psychological Safety Is the Hidden Engine Behind Innovation and Transformation," *Harvard Business Impact*, July 29, 2025, https://www.harvardbusiness.org/insight/why-psychological-safety-is-the-hidden-engine-behind-innovation-and-transformation/.
16. Carolyn Crist, "Nearly Half of CEOs Say Employees Are Resistant or Even Hostile to AI," HR Dive, June 4, 2025, https://www.hrdive.com/news/employers-employees-resistant-hostile-to-AI/749730/.
17. Andrea Childress, James Frazee, Brian Page, and Don Welch, "2025 EDUCAUSE Top 10 #8: Putting People First," EDUCAUSE Review, October 23, 2024, https://er.educause.edu/articles/2024/10/2025-educause-top-10-8-putting-people-first.

CHAPTER SIX

Campaigns and Revenue in the Age of Intelligence

Planning campaigns is about balancing inspiration through storytelling with a systematic execution of a plan. It takes us months, and for some, years, to craft a campaign narrative around the bold vision of the institution. The world is now moving much faster, and plans need to be more agile and dynamic than they were a decade ago. Donor attention, interests, priorities, and even financial capacity can change so fast, and we can no longer be built for data points updated annually or every other year.

Traditional campaigns were built as one-way broadcasts: We define the story, push it forward, and hope it resonates. An AI-era campaign functions more like a conversation, one that listens as much as it speaks. It analyzes live signals and adjusts accordingly, ideally daily. A gift officer compared it to leading a jazz ensemble, and it is now less like following a score and more like improv while the melody stays constant.

Annual fund teams are using AI to adjust messaging midstream during a giving day as open-rate and giving patterns emerge. Major gift teams are recalibrating portfolios when predictive models flag new prospects whose interests align with institutional priorities. Campaign communications departments adjust which stories they emphasize in donor media based on what content trends among alumni after a global event or a national day. These early signals point towards a new normal: Campaigns are becoming more like adaptive systems and less so like fixed plans in PDF documents.

Importantly, this shift is not about abandoning core fundraising principles. People still give to people, and the human connection still matters. Big ideas inspire transformational gifts, and consistent storytelling does matter. But in an information-rich environment, strategy must be responsive and adaptive. If traditional campaigns operated like road trips guided by folded paper maps with predictive routes plotted months in advance, an AI-enabled campaign functions more like GPS navigation. The destinations, our goals, are still the same, but the system continuously calculates, reroutes, and provides better recommendations based on traffic, new information, or weather. This chapter focuses on sequencing where to begin, how to pilot, how to scale responsibly, how to embed continuous improvement into campaign operations, how to design systems, not chase tools, and create meaningful donor experiences and more resilient revenue.

WHY THE CURRENT STRUCTURE FAILS AND HOW AI CAN CHANGE THE TEMPO

To design the future of campaigns, we first need to better understand why the prevailing model fails. The following steps are what we did for decades: conduct a feasibility study, identify top prospects, craft a case for support, organize leadership committees, set a multi-year plan, and then execute in carefully sequenced phases, starting with a silent leadership phase, followed by a public phase, then closing, then celebration. Yes, even in earlier decades, campaigns very rarely unfolded as planned. Over the last four decades, every major fundraising effort has been reshaped by forces that most likely were never part of their initial plans. The dot-com bubble of the 1990s, the terrorist attacks on 9/11, the 2008 financial crisis, the COVID pandemic, and, more recently, geopolitical volatility and accelerating technological change have altered donor expectations, priorities, and interests at speeds we have never seen before. Misalignment between when campaigns are conceived and when they are executed is structural and not situational.

Another structural challenge is the overreliance on the pyramid model. For those who are unfamiliar: This approach identifies the top 80 to 100 names of the pyramid, the donors who provide 60 to 90 percent of the campaign goal. With no labor to focus on the base of the pyramid, those opportunities are lost. AI lets fundraisers engage the bottom 1,000 names.

AI does not replace strategy, judgement, or relationships, but it does close the gap between donor behavior and the organizational response to it. Instead of depending on an annual portfolio refresh for gift officers, usually done in the summer, they can now receive a steady flow of signals that help them reprioritize portfolios weekly, or even daily. It might surface names in real time who are starting to behave like other major donors even though they have not donated yet, or flag long-time donors whose engagement patterns are slipping, giving the team a chance to intervene before their loyalty erodes.

If a recession looms, AI can help identify not only industries or regions with greater resilience but also counter-cyclical donors whose capacity might increase in downturns. Historically, it took months of pausing, observing, cultivating, and analyzing to understand which donors performed better during a specific crisis. During COVID, for instance, companies like Zoom didn't just weather the storm; they outperformed dramatically, yet advancement offices often realize this long after the fact, forcing a reactive "pivot." Today, AI simulations and predictive models can model that behavior ahead of time, allowing us to assess these scenarios before they happen.

This also applies to units within advancement. A campaign communications team can generate a story, and then, with help from generative AI tools, produce multiple versions of the story tailored to interest areas, life stages, regions, and donor histories, and see in real-time which one is doing well. This turns into campaign stories, with more living narratives evolving, rather than fixed scripts.

Organizations using AI-enabled tools report faster follow-up, more efficient travel planning, and substantial increases in donor engagement. These aren't marginal improvements; they signal a new campaign rhythm emerging.

Yet AI alone cannot solve structural problems. It is a means to an end. If a campaign plan is rigid, the process is too slow, or decision-making sits inside isolated teams, AI's insights will stall. We must redesign our campaign architecture, pacing, processes, team structures, and feedback loops to take advantage of what technology makes possible.

THE NARRATIVE INTELLIGENCE LOOP (DATA, STORY, ACTION, FEEDBACK)

The Narrative Intelligence Loop draws inspiration from iterative decision-making cycles used in fields like marketing and product design, but adapts them to a philanthropic context where story, not data alone, is the engine of action.

- **Data** is where the loop begins, not in the abstract, but in real signals donors produce every day and share with us and the world. It includes quantitative behavior (a gift, an event registration, a pattern of email openings), and qualitative cues (comments in a survey, a theme from a contact report, or a public professional update). These are treated as a steady feed, with a goal of turning this data into meaningful signals.
- **Story** is where creativity and strategic intent take over. Data alone cannot inspire a gift. It can, however, reveal what themes resonate, which needs feel urgent, and where donor values align with institutional priorities. The human review ensures the institutional voice and makes sure the narrative becomes both consistent and adaptable.
- **Action** is the deployment of that story across channels, like email, phone, text, proposals, events, or volunteer outreach. The key difference in the AI era is timing. Instead of sending messages on fixed calendars, teams respond to donor behavior as it happens and when it is still fresh in the donor's mind.
- **Feedback** closes the loop. Every action produces information: Did the donor attend, respond, give, decline, or offer feedback? Did the story resonate, fall flat, or end up trending? Did a new segment outperform? AI can summarize these patterns for us quickly, surfacing which messages convert, which need revision, and which donors require personal follow-up. We can continuously learn rather than conduct quarterly reviews of outcomes.

Case Study: William & Mary[1]

William & Mary (W&M), the second-oldest university in the United States, has become one of the most forward-looking institutions in advancement.[2] Its sustained success in alumni participation, eight consecutive years as the top public university in undergraduate alumni giving, didn't happen by accident. It emerged from a culture that values both relationship-centered fundraising and experimentation with technology.

Under the leadership of Vice President Matthew Lambert, W&M has embraced tools that strengthen human connection by automating the administrative work that often distracts from it.

Their partnership with GiveCampus illustrates what this blend of storytelling and systems thinking looks like in practice. In 2023, W&M became a design partner for the GiveCampus Gift Officer platform, which acts as an AI-enabled productivity assistant built to streamline time-consuming tasks like portfolio prioritization, trip planning, meeting preparation, contact reports, and drafting follow-up notes using a combination of generative AI, predictive AI, and agentic AI. After a month of using the tool, Lambert personally reported that he drafted more contact reports, completed follow-up emails more efficiently, and saved administrative time simply by letting the system do the first drafts based on his notes. More broadly, frontline fundraisers using GiveCampus scheduled an average of 2.2 additional visits per trip than before, leading to meaningful increases in engagement and a new pipeline.

The second major component of W&M's innovation ecosystem came through Givzey and its Version2.ai product. In 2024, Lambert was having dinner with Adam Martel, the founder of Givzey. Martel was the gift officer turned serial entrepreneur, co-founding Gravyty in 2015 with the intention of making fundraisers more efficient and effective by utilizing machine learning. This was not a new product from Martel's newest company. Martel shared over dinner that he intended to launch the first fully autonomous virtual engagement officer, similar to what was available at the time in sales, using AI-generated audio that mimics human voices with deep-fake videos that look almost real. The next morning, Martel presented this idea during the W&M advancement retreat, and they became a design partner with Version2.

By late 2024, W&M introduced "Wren," an AI-powered virtual engagement officer (VEO) assigned to a portfolio of several hundred mid-level donors, people who were too engaged for generic mass communication but too numerous to be managed individually by gift officers. Wren's purpose was to initiate and sustain personal-feeling dialogue via email, text, answering questions, sharing stories, and identifying who might be ready for deeper engagement. At the time, AI was capable in sales to generate human-sounding phone calls and fully interact with humans, mimicking human speech with "umm" when thinking or creating minor typos when typing phone messages.

This was not a small experiment; Wren and her VEO cohort across partner institutions took more than 18,000 autonomous donor actions in their first weeks and generated more than $150,000 in early gifts. These weren't transformational gifts, but the beginning of an experiment to solve for specific donor segments that are usually ignored or receive traditionally generic outreach. A human often stepped in when needed, or when the prospective donor asked for them.

The magic wasn't the novelty, but the orchestration. Wren was trained with W&M-specific messaging, voice, and values. She wasn't just sending emails but reinforcing the institution's narrative. And when donors expressed interest or concern, that feedback flowed back to the advancement team, where it shaped subsequent messaging and even influenced which campaign stories were elevated the following month. This is the Narrative Intelligence Loop in action: CEO engagement generating data, data shaping stories, and stories guiding the next action.

Of course, the work wasn't without challenges. Some donors embraced Wren's responsiveness, and others preferred a human touch and opted out. Transparency is always essential. W&M clearly disclosed that Wren was an AI assistant and not a staff member. The opt-out path was simple, and the advancement team consistently reviewed conversations to ensure tone, alignment with institutional values, and accuracy.

What W&M built and continues to build is not a futurizing fundraising experiment, but an early blueprint for what AI-informed campaign ecosystems will look like with AI tools that amplify human capacity, automated systems that reinforce narrative consistency, real-time engagement signals that feed back into the strategy, and ethical guardrails that work for the institution.

Case Study: Oregon State University Foundation[3]

If W&M shows us how AI can expand human reach, Mark Koenig, chief innovation officer and vice president for technology at Oregon State University Foundation (OSUF), shows how AI can reshape the architecture of a campaign. OSUF serves a large public research university with an equally large ambition: a $1.75 billion campaign already more than half achieved at launch. Their progress wasn't driven only by headline gifts; it was powered by an organizational shift toward integrated intelligence, cross-functional collaboration, and disciplined data governance. CEO Shawn Scoville understands and invests in technology capabilities, not just new vendor tools. The ever-evolving capabilities happened overnight and were very intentional.

The problems OSUF faced were typical of many of us, but their approach to solving them was unique. They built an Advancement Analytics and Technology unit that brought together CRM administrators, analysts, and data scientists together. This team wasn't buried inside IT; it sat at the strategy table. This helped turn analytics into a campaign engine, rather than being a reporting function only. The Advancement Analytics and Technology unit includes relationship insights and pipeline development, formally the unit of research, which is a very tired and old approach to information. By merging all of these systems together, OSUF became far more nimble and has increased the abilities of each one of these units. As Koenig said, "We forgot what IT stands for: Information Technology, not technology for technology's sake."

Take the example of OSUF's Women's Giving Circle (WGC).[4] For years, the program added roughly 10 new members annually. Using Fundmetric's predictive modeling, paired with an in-house model built by OSUF's data analysts, the team identified dozens of women donors whose engagement patterns matched those of past WGC supporters. With this insight, they launched a personalized outreach effort. The result: nearly 50 new members, an eleven-fold jump. AI did not just optimize strategy; it expanded possibilities. The special thing about their approach to predictive analytics is that they always want vendors to help them build internal capacity rather than simply signing up to sell their shiny new tool, product, or service. Vendor partners tend not to evolve with the work. Meeting the need in-house ensures innovative technology that keeps OSUF ahead of the curve rather than purchasing something outdated.

OSUF also invested early in AI stewardship. Recognizing that predictive models could unintentionally reinforce historical biases, they instituted routine bias checks. If the model's top prospects skewed towards a particular demographic, they examined the input and, if needed, retrained the model. They did not allow for black boxes (which come with privacy concerns).

Hearing their stories, their biggest innovation was not the data or the models, but the culture. They created a Digital Advocacy Council (DAC), a volunteer-led, cross-generational group of alumni and supporters who acted as campaign ambassadors. These volunteers did not just amplify messages; they helped co-create them, using analytics from OSUF to guide their outreach. When a lull appeared during a giving challenge in an engagement, the council proposed and executed a rapid response, be it a livestream thank-you message or a peer-led mini-match campaign. The DAC was the "training wheels" until OSUF could integrate the system into its daily life as a data-driven organization.

OSUF also introduced job descriptions 2.0, incorporating AI literacy, adjusting expectations on every level of the organization, and showing that AI is an important skill that is needed, not just a tool to experiment with when convenient.

OSUF's case shows that intelligent campaigns don't have to begin with AI tools alone but can begin with integrated systems and a culture willing to collaborate. Anyone doing AI work for technology's sake is the reason we will see a bubble pop this year or next. Then we will see the real winners and how to proceed.

What makes OSUF even more instructive is how intentionally they built the surrounding systems, structures, and habits that allowed innovation to take root. Their progress came from a decade-long commitment to planning, clarity, and disciplined iteration.

Beginning in 2016, OSUF mapped a clear, staged evolution of its AI and automation maturity. They modernized the CRM, partnered early with Fundmetric, cleaned and restructured data, launched predictive models, introduced smart gift forms, piloted AI-generated stewardship videos, hired a senior director for automation and change management, created an in-house data scientist position, and built automated pipelines and segmentation engines. Each step was small, but the cumulative effect was transformational.

Each vacancy was treated as an opportunity to reimagine roles, and the team revisited job functions every two to three years to match the pace of technological change. Staff development seems to be treated as a strategy and not a perk. Culturally, OSUF is embracing an innovative mindset grounded in curiosity and psychological safety.

Finally, OSUF adopted what they call the Symphony Approach: a model where innovation is orchestrated across the enterprise rather than isolated in one corner. The Technology & Innovation Division set the strategy, the analytics team generated insight, the automation teams activated it, the Digital Advocacy Council amplified it to the community, and frontline fundraisers brought it to life through human relationships. Innovation at OSUF wasn't a department; it was a culture aligned around shared purpose.

Case Study: Clark University[5]

Clark demonstrates how a smaller advancement shop can combine clarity, focus, and design with fewer staff and a smaller budget than the above two cases. With only 30 FTEs, Clark could not rely on the sheer volume of activity. Instead, a campaign engine was evolving, defined by precision: the right signal, the right moment, connected to the right opportunity. The campaign is being led by George Bacher, senior executive director of advancement strategy and campaign planning.

The Four Data Frontiers from chapter two are being scaled, the data gathered and analyzed to map out the best first draft of donor strategies.

Instead of constructing a full intelligence ecosystem all at once, Clark is piloting a series of tightly scoped, high-impact tools designed to make campaign strategy more accessible, more adaptive, and more consistent.

At the center of this work is an emerging AI-enabled campaign toolkit. Rather than treating campaign materials as static PDF documents or website sections, Clark is testing a prompt-driven interface where fundraisers and key volunteers can interact with campaign content or draft talking points for specific scholarships or turn a long report into an AI-generated, human-sounding podcast. This toolkit is aimed at reducing friction, eliminating guesswork, and giving staff faster access to the story they need.

Clark is also experimenting with AI-generated first drafts of briefing packets, historically reserved for top senior leadership due to the time and human hours previously needed to generate them. These packets synthesize and make available donor history, institutional priorities, relevant news, suggested meeting objectives, and potential next steps in minutes to all fundraisers and to virtually any prospect in the portfolio, turning this into a feature provided to 1,000s and not just the top 50 to 100 names. Fundraisers still refine the content, providing the human touch, the editing, and the fact-checking, rather than compiling, which often takes hours.

At the same time, Clark is exploring scenario-based predictive modeling to help leaders rehearse possible futures before they materialize: What if a recession compresses giving in certain sectors but expands it in others? What if national policy impacts giving negatively or positively in a certain priority? By running small simulations against donor segments and engagement patterns, leaders can predict how to adjust portfolios accordingly.

Using Signal by EverTrue and its companion, DonorSearch.ai, frontline fundraisers can receive signals daily and act on them, using the AI assistant to chat with a donor record, asking it questions, or talking to the database and using data points available from DonorSearch.ai and its predictive modeling functions.

The advancement communications team uses AI to generate first drafts and as an idea-generating companion. The campaign website has been updated to provide a traditional browsing option or a prompting option for specific requests. If a donor wanted to give $100K in real estate to support student scholarships, the campaign website could respond to this request, letting them know from the Gift Acceptance and Advancement Policy that Clark accepts this type of gift, sharing a donor story from a similar type of donor, including a video from a student scholarship recipient, and offering to connect them with the appropriate advancement colleague.

Campaigns succeed when they operate as adaptive systems. The next section explores what the system looks like and how campaigns can shift from a static pyramid to a continuous loop of engagement supported by agile teams, real-time data, and coordinated decision-making.

CAMPAIGN COMMAND CENTERS COORDINATING CONTINUOUS ENGAGEMENT LOOPS

As campaigns move towards continuous engagement, many institutions are adopting what can be best described as campaign command centers: a coordinated operating rhythm that brings data, strategy, and action together. They are designed to give teams the same real-time situational awareness found in political campaigns, modern social marketing campaigns, and emergency operations, fields that have long recognized the value of rapid coordination.

These command centers are not meant to replace existing units like annual giving or advancement communications, but instead connect them to create a shared vantage point from which teams can see what is happening, why it matters, and who should act next.

The inclusion of agile workflows also reflects practices long established in product development, software engineering, and DevOps. But these have rarely been translated into the architecture of a multi-year nonprofit fundraising campaign.

Finally, the emphasis on cross-functional decision-making draws from the operating models used in management consulting, tech firms, and healthcare coordination teams. I know sometimes we struggle with barely making decisions made in one functional area, but this is where we are heading, and the faster we build these systems, the faster our culture will change to adapt.

A command center often includes several core elements:

1. A shared real-time dashboard that acts as a single source of truth.
2. A cross-functional meeting cadence is most commonly weekly and, in some instances, daily, to make small but meaningful adjustments to segments, outreach, stories, or who needs to coordinate follow-up.
3. Rapid-response decision protocols to empower teams to act quickly within predefined boundaries that won't require decisions to wait for managers or committee meeting cycles.
4. Integrated content and messaging from a central repository of content or an AI-assisted content library to ensure that each unit is telling the same story in a way that can be tailored for different audiences.
5. Clear escalation and collaboration. This avoids the traditional campaign problems of opportunities stalling between one-on-one meetings, shared inboxes, or falling between departmental boundaries.

The core value is not the space itself, but this command center can turn into a physical space with live feeds and reports. Our advice is to focus on coordination, not space. This approach has the potential, if deployed properly, to turn the team into more proactive, less siloed, and more confident in their ability to adapt. Momentum becomes easier to maintain as intelligence is shared and not scattered.

HOW CONTINUOUS ENGAGEMENT TRANSFORMS THE DONOR JOURNEY

This type of campaign is most likely to make donors feel more engaged as they experience a campaign that listens, responds, and adapts to their interests and signals.

1. Donors will experience relevance, not repetition. They no longer get appeals that ignore their last interaction, and they receive invitations that reflect it.
2. Stewardship becomes timelier and more meaningful.
3. Donors see themselves in the campaign narrative. The stories become more of a mirror that best reflects the donor's interests and what they care about the most.
4. Momentum builds earlier and more naturally, activating the middle and long tail of the donor base earlier, and momentum does not have to wait for a public launch, but grows organically through thousands of micro-engagements.
5. Predictive insight guides the next best move, not just what our gut tells us based on our available time.
6. The campaign's optimal performance is when it evolves with donors and their interests over time.

A CAUTIONARY TALE: WHEN AUTOMATION UNDERMINES TRUST

Even with strong guardrails, missteps can happen. And when they do, they often happen fast. Later in the book, we will hear about the exceptional work of Vanderbilt University as a pioneer in the ethical and innovative use of AI. What follows here is not a critique of their strategy, but a reminder that even institutions doing nearly everything right can experience a simple, small error with outsized impact.

In 2023, after a tragic shooting at Michigan State University (MSU), Vanderbilt's Peabody College sent a condolence email to students in solidarity and to express their empathy for the events at MSU.[6] At the bottom, a disclosure said, "paraphrased from OpenAI's ChatGPT." This kind of transparent disclosure is important, but for something this serious, and for an email only a few paragraphs long, the community expected a human to make the effort and write it themselves. Students felt that a message sent to them with the intention to provide comfort suddenly felt algorithmic and generic.

It bears repeating that Vanderbilt is an early leader in institutional AI innovation, and perhaps dozens of its use cases serve as national models. This moment was an exception but had an outsized impact on the reputation. The lessons here include:

1. High-emotion moments require a human-only approach.
2. Transparency is important, but pairing it with human judgment is necessary.
3. Automation is powerful but unaware. The GPT did not alert them to this sensitivity and kept filling in the next word of the letter.
4. Small mistakes can have national consequences for the institution.

12-STEP ROADMAP TO AN AI-INFORMED CAMPAIGN ENGINE

Step 1: Establish the Vision and Secure Leadership Buy-In

Every intelligent campaign begins with clarity of purpose. Before selecting the tools or designing the pilots, leaders must articulate why an organization is pursuing an AI-informed approach and what outcomes matter most. This is more of an AI mission statement and not a technology position statement. The statement should be able to answer questions like:

Can AI help us strengthen donor relations? If so, how? Can it help us tell clearer and more timely stories? Increase strategic capacity, not just efficiency? Uphold our institution's values and ethical commitments?

Ron Shaich explores this using the ends, means, and byproducts framework:[7]

- The **end** is deeper engagement and a more meaningful relationship with the institution and its constituents.
- The **means** are intelligence workflows, real-time insights, and data-informed storytelling.
- The **byproducts** are better donor experiences, higher staff capacity, and more resilient pipelines.

Leadership buy-in is essential, and without it, hesitation and fear will form within the team. When presidents, provosts, boards, and vice presidents understand purpose, they become amplifiers of the work, not gatekeepers. Clarity of the vision reduces fear and helps staff better embrace change.

Step 2: Form the Cross-Functional AI Task Force

AI-informed campaigns cannot be built by one team or one unit alone. They require coordination across roles that historically operate separately.

The AI task force should be:

- Cross-functional, representing all major advancement areas.
- Empowered not just by mandate but by recommending and executing pilots and sequence decisions.
- Accountable, with clear scope, cadences for meetings, and deliverables.
- Trusted and visible to leadership and transparent to the rest of the advancement staff.

This group becomes an internal innovation spine and not a technology committee. Its role is to identify high-impact use cases, ensure ethical review and data governance, align pilots with campaign priorities, keep the work pace, and communicate progress across the organization. Make sure to include skeptics in the task force so the discussion does not evolve into an echo chamber. At the early stage, the task force does not need a very complex charter; simplicity at that stage is key: a mandate from leadership, a shared understanding of the vision, and a commitment to move deliberately and not chaotically.

Step 3: Audit Your Data and Map the Four Frontiers

Before building any intelligent campaign system, you must understand what data you have, where it lives, how reliable it is, and who maintains it. The goal is not perfection but finding patterns of which frontiers are rich and where blind spots exist.

Accurate data is the fuel for intelligent campaigns. Mapping the frontiers ensures the institution knows what fuel it has and what must be strengthened before automation, triggers, or modeling begins.

Step 4: Clean, Connect, and Establish the Data Backbone

Once the data is mapped, the next step is to learn and connect it. You should resist the temptation to "boil the ocean." This step includes establishing a data governance baseline: Who approves new fields? What expires and when? Who maintains data hygiene? What checks prevent misuse and bias? And what can be collected, and why?

Step 5: Launch a Small, High-Value Pilot (Start Simple and Focused)

With data mapped and connected, the next step is to run a small, tightly scoped pilot that demonstrates what intelligence can do. If you are working with a vendor, they might be able to run a proof of concept as a pre-pilot that helps you test the value of the pilot with minimal time commitment.

The best pilot shares these characteristics: (1) They solve a real problem, (2) they cross at least two data frontiers, and (3) they produce observable outcomes quickly. Once you operationalize the pilot, the wins become real.

Step 6: Train and Upskill the Team (AI Literacy as Campaign Capability)

This step is about ensuring that everyone has baseline AI literacy and that the team has a strong foundation in verticals related to their specific roles. You're not trying to convert them to data scientists, but rather teach them about role-based learning, so if they're given predictive signals, your team knows what the signals mean, what they don't mean, and how to use them.

Providing hands-on practice and providing ethical and responsible use training for everyone in your team is important. This step also includes creating continuous learning cycles, so the learning is not a one-off but includes monthly refreshers and office hours as needed. The best shops include sessions for team members to share what they learned at work and outside of it on a regular basis.

This step is not a technology class, but it is meant to create a culture that is comfortable working with both human and machine intelligence side by side.

Step 7: Expand What Works

Not every pilot will succeed. Embrace that and learn to be comfortable pulling the plug on a project. This step is about understanding what works, why it works, and how to scale those elements thoughtfully across the campaign. Key principles of scaling include scaling outcomes and not the tools, letting evidence and data guide your expansion or growth, and not scaling too soon, as it would drastically reduce chances of success. Most importantly, keep humans in the loop and document the process. In the future, these notes will help see patterns for other pilots, tools, and initiatives.

Step 8: Implement the Narrative Intelligence Loop

This is where the campaign starts to become truly adaptive. The loop is the operating model that moves campaigns from static plans to living systems.

Step 9: Establish Ethical Guardrails and Governance

By this stage, the campaign is gaining intelligence and speed. This step ensures that speed never comes at the expense of trust. Ensure that you use your Ethical AI Use Policy, the Intentional Legitimate Interest Assessment tool, and the other steps we discussed in chapter two.

Step 10: Engage Donors with Transparency and Agency

Continuous campaigns succeed when donors feel respected, informed, and empowered, not "targeted." This step is about ensuring that the donor experience remains the center of every intelligent workflow. The foundational practices include:

1. Transparent communication norms.
2. Clear preference management.
3. Respect for donor signals and what they truly mean.
4. Donor choice in engagement journeys.
5. Honest storytelling anchored in data, impact, and lived experiences, not hype.
6. No algorithmic "black boxes"; each system should be understood before it is used.

Transparency converts personalization from something that feels manipulative into something that is thoughtful.

Step 11: Launch the Continuous Campaign Model (Move from Moments to Momentum)

This is not marked by a dramatic unveiling but by new rhythms, habits, and coordination patterns becoming the way the shop works. Key elements of launching the continuous model:

1. Activate the campaign command center.
2. Treat strategy documents as living frameworks.
3. Leverage predictive insights for prioritization.
4. Maintain the cadence of donor-centric responsiveness. These timely micro moments have the power to convert signals into trust and trust into action, including giving.

This shift in posture is likely to happen; be thoughtful about it, but don't expect a "go-live date" for it.

Step 12: Make Intelligence a Habit: Measure, Iterate, and Sustain

The final and perhaps the most important step in this process is protecting the loop. Your campaign must be constantly learning, evolving, and renewing. What you worked hard to establish in a year or two might be obsolete or no longer needed by your key constituents, and that is fine. Learn, evolve, and iterate. To succeed in effective iteration, you must clearly define what success means to you beyond dollars raised. Define "success metrics," review insights regularly, and perform bias audits frequently.

We strongly believe that people give to people, and as campaigns evolve, the ones that will inspire the largest gifts of a lifetime will always be a human-to-human endeavor. AI is not the hero of this chapter, but rather a means to a new set of capabilities that expand the reach and relevance of our mission.

Chapter Six Recap

Core Message

In the AI era, campaigns are adaptive and human-centered, combining insights and ethical stewardship.

Key Takeaways

- Continuous campaigns replace episodic campaigns.
- Command-center coordination creates coherence.
- Ethics project speed and trust.

- AI enhances human work and, at least in relationship building, should not replace it.
- Continuous learning to keep up with change is a pathway to sustaining campaign momentum.

Why It Matters

The future of mission-driven organizations is about how to be adaptive, attentive, intelligent, and anchored in trust.

Chapter Endnotes

1. Matthew Lambert (vice president for university advancement, William & Mary), interview by Joe Manok and David Woodruff, 2025.
2. "About W&M," William & Mary, accessed March 2, 2026, https://www.wm.edu/about/history/.
3. Mark Koenig (chief innovation officer and vice president for technology and data, Oregon State University Foundation), interview by Joe Manok and David Woodruff, 2025.
4. Rachel Crosbie, "Big Impact, Small Data: The 11x Growth Story in Annual Giving," Fundmetric, accessed March 2, 2026, https://www.fundmetric.com/blog/osuf-wgc.
5. George Bacher (senior executive director of advancement strategy and campaign planning, Clark University), interview by Joe Manok and David Woodruff, 2025.

6. Rachael Perrotta, "Peabody EDI Office Responds to MSU Shooting with Email Written Using ChatGPT," *The Vanderbilt Hustler*, updated February 21, 2023, https://vanderbilthustler.com/2023/02/17/peabody-edi-office-responds-to-msu-shooting-with-email-written-using-chatgpt/.
7. Ron Shaich, *Know What Matters: Lessons from a Lifetime of Transformations* (Harvard Business Review Press, 2023).

CHAPTER SEVEN

Beyond Advancement: AI Across the Institution

At many mission-driven institutions, AI is cropping up everywhere. Marketing departments tap into generative AI for social media content, admissions offices deploy chatbots to handle common applicant questions, career centers experiment with resume review agents, student success teams use early-alert algorithms and advising bots, and academic staff explore AI research assistants. The tools have arrived, but on most campuses, they are siloed. One department's chatbot knows nothing of another's automated tutor, and data practices remain fragmented. The lack of connective tissue creates an operational headache at the institutional level. When AI adoption is uncoordinated, it poses a serious set of challenges. According to *Inside Higher Ed*, some units adopt different tools for similar needs, leading to "confusion, duplication, and hidden costs."[1] *Inside Higher Ed* highlights that isolated tools without shared standards can cause each AI system to develop a different "voice." One student might get an upbeat, casual tone from an admissions bot while another gets a formal, scripted tone from financial aid. A consistent experience is crucial.

Instead of workers feeling empowered, many are overloaded. Unchecked fragmentation also heightens this feeling. When you implement AI with this level of departmental separation, it likely means the guiding principles or governance are not yet set. When this lack of coordination happens, possible compliance-related issues are also likely to surface, including inconsistent ethics and quality control. One key byproduct of self-policing AI is becoming vulnerable to misusing AI in ways that damage trust. This is particularly dangerous for nonprofits, universities, and other fundraisers. To truly harness AI for mission impact, nonprofits and universities must treat it not as a collection of apps, but as an institution-wide infrastructure challenge. The stakes are high, and done right, campus-wide AI adoption can boost efficiency, personalization, and overall mission delivery.

AI AS INFRASTRUCTURE, NOT APPS

So, how do we become an AI-ready organization? We believe the answer to that involves thinking of AI as an infrastructure, part of your organization chart, not just applications or scattered tools that you subscribe to here and there. This would require creating shared services of AI models and AI agents that any unit or department can use. Second, establishing standards for prompts and data use. Amplify even allows users to create prompts that colleagues can use and share with each other.[2] Unified data governance is another pillar. AI readiness is demanding a cross-domain data policy covering student records (FERPA), health information (HIPAA), donor and client data, and other sensitive information. Without a unified policy, the organization faces risks: One department might load a spreadsheet of student grades into an AI tool, resulting in the violation of privacy rules that are being carefully followed by another department. An infrastructure approach sets clear guidelines on what and how that data can be used with AI. These should be part of thoughtful negotiations with vendors and AI platforms.

Consider this in a nonprofit setting. Staff could use a central AI assistant to analyze their program data, confident that the data won't leak out or violate any client confidentiality policies set by the organization. Just as a campus network has access controls, firewalls, and a virtual private network (VPN), AI needs ethics and oversight mechanisms embedded in its system. This might include content filters that are automatically applied or warning signs when something is being asked or about to be uploaded that could violate policy, ethical standards, or the law. Some universities have formed AI ethics committees and review boards that evaluate proposed AI tools for compliance with their regulations and values. At Clark University, AI review was incorporated into the existing IT Governance process, where any new technology-related product or service was already being evaluated to incorporate the appropriate use of AI.[3] For instance, the central AI platform can have a feature that logs the metadata of the interactions for auditing purposes, much as they do with other enterprise-level systems. In this way, it would become feasible to conduct periodic reviews with the ability to answer questions like: Are AI tools being used as intended? Are there any red flags or loopholes in the policies?

In 2025, an AWS article described how many campuses are stuck in AI "pilot purgatory."[4] Similarly, a Civitas Learning report warned that disconnected systems and teams lead to wasted resources and missed opportunities.[5] This makes AI adoption as seamless as Wi-Fi or email adoption at a university or a nonprofit. Soon it'll become a part of everyday work.

Earlier in chapter six, we saw Matthew Letendre from Dana-Farber Cancer Institute using their centralized AI tools to do his research and work from a fundraising standpoint. Such deployment is far safer and in compliance with donor privacy and ethical standards, local laws in Massachusetts, and federal laws.

Case Study: Vanderbilt University's Amplify Sandbox[6]

How does an institution build connective tissues for AI?

At Vanderbilt, this became a real-world answer when, in 2024, they launched Amplify, their custom GenAI sandbox designed to be used campus-wide by faculty, staff, and students.[7,8] The initiative was born directly out of the need to coordinate AI innovation at a time when GenAI hype was at its peak. Amplify offered a do-it-together approach, where a single platform provided a safe-to-fail environment with guardrails built in to support its users and keep all the data within the university's control.

Vanderbilt's Amplify is a personalized version of ChatGPT that runs on their private cloud using AWS. Users log in with their university credentials and engage with different large language models, but to the users, it feels like one AI assistant. Unlike public AI tools, this version is more integrated and promotes trust. A faculty member can use it to brainstorm lecture materials; a staff member to draft an email, summarize a report, or analyze survey data; and a student can use it to get feedback on a project idea. This enterprise-level coordination allows for successes and learnings to be shared easily across "silos." Vanderbilt augmented their platform with training modules to guide users and support channels to make sure they are increasing their chance of success.

The Amplify governance model is collaborative. It was funded by the provost's office with support from the chancellor and the central AI unit (VUIT). This ensured buy-in from both the administrative and the academic leadership. A dedicated technical team led by CTO Allen Karns built and maintained the system, but also works closely with stakeholders across the university to identify needs and share outcomes. Amplify became interwoven into the university's strategic plan rather than an isolated one-off project of its own. The other firepower behind the project is computer science professor Jules White, who, at the time of writing, served as senior advisor to the chancellor on generative AI in enterprise and education. The project is now housed in the Amplify GenAI Center of the College of Connected Computing.

One notable practice is their pilot-to-production pipeline for AI ideas. Amplify was a pilot itself. This helped with making the design process more inclusive of users and allowed time for gathering and ironing out technical issues and challenges. Once the campus launched widely, the ethos of piloting continued at the use-case level. The university invited faculty and staff to propose AI projects that could leverage Amplify. They expected a handful of submissions and were pleasantly surprised by the 180 proposals submitted that spanned from automating routine reports to creating AI study aids. This bottom-up energy was channeled by pairing selected "Staff Fellows" who submitted the ideas with both AI developers and student helpers to build out the ideas. Essentially, Amplify became the backbone series of a sandboxed set of pilots: Try an AI-driven idea in one department, evaluate it, then scale it if it is successful across the institution as applicable.

Some of the examples of what emerged from this sandbox approach included: a staff-built PowerPoint generator that automatically creates university-branded slides from a text outline, saving hours for many employees. Another team connected Amplify to their Microsoft Outlook system to generate nightly email and calendar summaries for staff, an amazing productivity tool. These might look like small wins, but collectively, they are removing friction at Vanderbilt for day-to-day operations and helping create momentum for people to see tangible value and potentially the bigger picture of why these technologies are important and how they can help them in real time.

This approach also closes the feedback loop between the pilot and production. If a particular AI use case does not work out, whether it does not actually save time or the quality of the output is not up to user expectations, the sandbox environment makes it relatively easy to learn from that experience and move on without any major sunk costs, time commitments, or operational disruption.

Conversely, when something works out, Amplify provides pathways to scale it campus-wide or across multiple departments. An illustrative success loop is how Amplify embedded this into teaching and research: Early adopters among faculty experimented with AI in course preparation, reported results to colleagues through workshops across the university, and soon more faculty started integrating it into their workflow. By late 2024, the university felt confident enough to extend access to Amplify to all students alongside guidance on academic integrity.

Had each faculty used separate AI tools, which is the case in most academic institutions at the time of this book, that rollout would've been siloed and chaotic. Depending on a unified platform centrally managed, Vanderbilt could address concerns like bias and data privacy before opening the gates to students.

Of course, this is also about culture and not just technology, as is always the case with deployment and adoption. Vanderbilt's leadership, all the way up to the chancellor, actively championed AI experimentation, signaling that thoughtful innovation does not fear progress. This top-level endorsement and support meant that when Amplify launched, it was not met with the usual institutional inertia or turf wars. Instead, faculty and staff were already primed to ask, "How can I use this in my area of work, research, or teaching?" rather than asking, "Why is central IT imposing tools on me that I do not want to use or like or need?" Put simply, Amplify did both sandboxing the technology and engaging the community. It turned AI from a buzzword into a collaborative problem-solving mechanism across departments.

Finally, Vanderbilt's model emphasizes ethics and value alignment as part of their model. The university fosters a holistic approach to responsible AI through the launch of the new College of Connected Computing, which, with its "Computing for All" ethos, will address concerns about the inclusion of AI across all disciplines. By baking these conversations into Amplify's rollout, the university helped ensure that as AI use scales, it does so in line with the university's inclusive, student-centered mission.

Amplify's success at Vanderbilt shows that institutional orchestration of AI is not only achievable but vastly beneficial. In roughly a year, they went from a scattered web of experiments into an integrated, platform-enabling ecosystem. The key to their success is a central, flexible infrastructure, with clear data and ethics guardrails, leadership endorsement, and programs to involve people in the experimentation.

Case Study: Amazon Web Services (AWS): A Cross-Sector AI Transformation View

Zooming out beyond a single campus or a single deployment, how are larger ecosystems approaching AI integration and adoption? Few organizations have a broader view of cross-sector AI transformation than Amazon Web Services. They work with thousands of universities, nonprofits, and government agencies on cloud and AI projects, and they have become a thought leader in guiding these institutions into the AI era. Their work aligns with what we have seen so far: a focus on orchestration and architecture, not on hype. AWS began working on Agentic AI as a model for AI enablement. They emphasized AI systems that can orchestrate actions and workflows, not just make predictions and chat.[9]

One way AWS frames this is by encouraging organizations to see AI along a continuum of capability: from basic tools to enhanced workflows to transformative system change. Maysam Ali, who leads the Global Education go-to-market at AWS, notes that many institutions start by piloting individual AI tools (say an admissions chatbot or a fraud detection ML model). That works, but the real impact comes when you connect these components into end-to-end solutions and eventually reimagine processes. She referred to this as a layered AI transformation. For instance, a university might first adopt a generative AI tool to auto-translate outreach materials. Next, they might integrate that into a workflow: An AI translates, then automatically routes the content for review and publishing. Finally, the university might redesign its entire content strategy around instant multilingual output, teaching communities faster.

AWS team members frequently urge users not to stop at the pilot and then scale it, but to connect it to their mission pipeline. In higher ed, this could mean linking an AI tutor pilot to the tutoring center's operations and eventually to curriculum design, rather than treating it as a one-off app.

Central to AWS's approach is viewing AI as an architectural discipline. Just as cloud computing requires architects to design elastic, scalable systems, AI requires thinking about AI engineering: model management, integration patterns, data pipelines, and governance structures. AWS has formalized some of this in its products, notably Amazon Bedrock, a service used to build and scale generative AI applications.[10] In 2025, AWS introduced features like Amazon Bedrock AgentCore to help organizations deploy "production-ready AI agents at scale" with the needed security and observability.

The technical details are not as important as the philosophy here. An AWS blog on higher education put it this way: Universities are eager for agentic AI, but they get stuck in "pilot purgatory" without the right integration infrastructure.

The solution is purpose-built components for things like connecting AI agents to internal systems. For instance, a gateway that links an AI agent to the student information system and monitors the interaction to audit the agent, step by step. AWS provides such components, but more importantly, it educates leaders that "these architectural pieces must be in place" if you want secure, scalable AI. It is not just spinning up a chatbot and hoping for the best. During AWS's 2025 Nonprofit Generative AI Week, one speaker described generative AI as giving them "a virtual set of extra people that you didn't know you had."[11] You must consider these virtual colleagues as actual colleagues who require onboarding and regular training. AWS showcased how nonprofits are translating this concept into practice:

Worldreader, for example, used generative AI to translate an entire digital library into multiple languages and create personalized reading suggestions for families. The tech (a large language model) was just one piece; they had to integrate it with their content management system and align it with educational goals.

Another nonprofit built an AI chatbot (Myelo) to answer complex questions from cancer patients, which handled 20,000 questions in its early run. Success for them meant that the chatbot was orchestrated with their knowledge base and had guardrails for medical advice. This was part of the service infrastructure and not a gimmick.

On campuses, AWS sees similar patterns. Universities are deploying student support agents who can guide a student through financial aid forms at 2 a.m. or answer questions about campus resources. But the value multiplies when those agents are connected to back-end systems that pull the student's specific data, flagging missed classes and proactively scheduling appointments on an advisor's calendar to support student success.

AWS also encourages a "micro-innovation" approach akin to Vanderbilt's staff fellows to avoid analysis paralysis. Patrick Frontiera, who leads AWS's US Higher Education Strategy, noted that sometimes simple automations yield big wins.[12] Take the Highline College's AWS-powered financial aid status tracker that cut inquiries by 75 percent.[13] By funding or greenlighting small departmental AI experiments through micro-grants and pilot programs, institutions can build literacy and proof points. AWS often partners in these pilots to provide credits or technical support, with the understanding that if it works, the institution will scale the AWS pilot and connect it to enterprise systems. This approach dovetails with AWS's emphasis on cross-departmental literacy. They frequently host multi-stakeholder workshops bringing together faculty, administrative leaders, and IT staff to jointly explore AI use cases that support our mission. They "work backward" from the customer problem and what they're trying to achieve, then look at how technology can help address that problem. The idea is to break down silos early, with everyone grasping the basics of AI and the organization's strategy for it. AWS also offers AI literacy programs for its staff that are open to the public, like the AI practitioner certification.

Finally, AWS is helping organizations bake ethics and governance into the pipeline. They've published a Generative AI Best Practices Framework and provide tools like model audit logs, bias detection, and human review workflows on their platform.

Expert Voice: Maysam Ali

Amazon Web Services on Agentic AI[14]

In my work across education, I often get asked: *What are the best uses of AI, and how can we help our staff, faculty, and students develop AI literacy?* This is top of mind for leaders, as the technology is fast-moving and there is an imperative to provide the tools, training programs, and best practices to evolve with it. The rise of agentic AI has expanded what's possible for organizations. We're moving from AI that provides quick answers toward AI that can act on your behalf, and that is a far more powerful technology.

When I say agentic AI, I mean AI systems that are endowed with a degree of autonomy where they can plan, decide, and execute tasks to meet a specific goal that you have set for them without needing you to prompt every step. Think of the difference like this: Generative AI might help draft five email variations when you ask it to do so. Agentic AI, by contrast, could be told to increase our donor re-engagement in any given month, and it would proactively draft emails, send them out at optimal times, maybe even segment the recipients, and then report back on results. Agentic systems can pull data from across systems, analyze it, and execute multi-step workflows to achieve a goal.

Take the process of admissions at any university. According to the NASFAA 2025 Administrative Burden Survey, 91 percent of financial aid offices report an increased amount of time and resources devoted to processing each application over the past five years.[15] More than half of financial aid administrators say their offices experienced resource shortages during peak periods over the past five years. For international applicants, there are additional challenges, with documents arriving in dozens of formats and languages, and grades following unfamiliar scales. Agentic AI enables review across formats and languages, conversions of grading systems, and integration with CRM platforms to accelerate enrollment decisions. This speeds up this process and ultimately delivers an improved experience for students as well as administrators. This is another example of an agent orchestrating a complex, multi-step and end-to-end process.

This autonomy is powerful, but it's also why we must design these systems with responsible practices from day one and deploy guardrails that support AI safety. This includes defining the level of autonomy, validation loops, and human oversight. Thinking of the human world, we give new employees job descriptions, access permissions, and manuals or standard operating procedures. You can think of a similar structure for AI agents. When we deploy an agent to handle some donor communications, we need to contain its scope: For example, it should only email donors below a certain threshold, it should only use approved messaging, and it should never transfer money or change donor records by itself. The boundaries you set involve sandboxing the agent's access using role-based credentials, as we do on AWS, so the agent literally cannot perform actions outside its allowed domain. We also set time limits or resource limits. Autonomy does not mean carte blanche. Leaders must ask what walls to put up so this agent operates safely within its intended zone.

Next is coordination. In any complex organization, you won't have just one agent doing one thing, but rather a combination of multiple agents and humans, all interdependent. Picture a future student services scenario with one AI agent nudging students about deadlines, another scheduling their tutoring sessions, a human counselor handling the sensitive cases, and maybe a third AI agent monitoring for risk patterns.

These entities need to work in concert with each other, and without that level of coordination, you get chaos in the form of duplicative or conflicting messaging. So, we built a coordination layer for that reason. Practically, this could mean a centralized orchestration system where agents register their tasks, or a communication protocol that agents use to communicate with each other and with human supervisors. One emerging solution we use is having agents use something called the Model Context Protocol to interface with various tools and with each other. Technical details aside, the principle here is that we design how agents fit into workflows. Who hands off to whom? And how do you prevent an AI-vs-AI collision? Like what we do with humans and teams anyway.

Finally, oversight. Oversight means having the ability to understand, audit, and, if needed, intervene in the agent's behavior. In human teams, we have managers, we have auditors, and we have retrospectives on projects. For AI agents, we do the same so that when something anomalous happens, the agent can try the same action multiple times or produce outputs that trigger a flag. One best practice is to have a 90-day agent audit so that every quarter you review the agent's logs and performance and decide to continue, adjust, or decommission.

The good news is that AI can orchestrate this. For example, at AWS, the team built observability dashboards that allow us to see each step the agent takes, like a flowchart, and we can pinpoint where it potentially goes off track. Leaders should insist on that level of transparency before allowing an agent wide latitude.

There is also a cultural aspect of this, related to concerns about job security or overall disenfranchisement with agents. Frontline operators might either over-trust the AI or feel threatened by it. Leaders must set the tone: They must communicate why the organization is adopting AI. In most cases, it is to augment human capability, not replace it. Policies and job descriptions help people understand how AI proficiency and workflows fit into their jobs. Just this month, I will be sitting for the AI certification at AWS to ensure that I have certain knowledge and skills to do my job, given where technology is now and where it is heading. We do this on an ongoing basis, and I strongly encourage anyone with curiosity about this technology to do so. The best deployment is when your team sees AI as a colleague there to support their work, free them of tasks that may be repetitive, and allow them to focus on areas that they would like to focus on.

To go back to the advancement context, agent collaboration is critical at an institutional level. AI agents working on multiple projects collaboratively can reduce friction between student success, admissions, communications, advancement, and the registrar on things that really are operational and do not require much friction in the first place. That level of AI deployment at the institutional level is where most meetings, emails, and tension points occur. Automating that allows your humans to focus on what matters most.

Toolkit: Designing Connective Institutional AI Capacity

1. <u>Shared Prompt Library</u>

   One of the most important things to demystify is how to ask X questions, as many users will resist using AI simply because they do not know how to ask. The first step is to identify a list of prompts that can be most helpful for the majority of your users.

2. <u>Unified Policies</u>

   Create a cross-functional data governance policy, and add to it specifics for each department or unit based on their needs and uses. You can choose to classify data sensitivity and what your community can use or should not use with AI tools.

3. <u>AI Sandbox</u>

   Create an AI sandbox where all your internal stakeholders can connect with a single sign-on and help you preload AI models that you prefer to be used.

4. <u>Internal AI Literacy and Certification</u>

   Invest in building AI fluency across all roles in your institution. Some create a basic AI literacy certificate for staff, while others have more advanced credentials, like AWS, where they have a web of functional and technical certifications, like the AWS Certified AI Practitioner.

5. <u>Agent Lifecycle Management Template</u>

   Prepare for an era of AI agents by instituting a lifecycle checklist for any semi-autonomous system that you will establish, kind of like when you hire someone on a contract for X years to perform Y duties and be managed by Z colleague.

6. **Cross-Functional AI Innovation Fund**

   It's key to create a fund that enables piloting, scaling up, and cross-disciplinary work, as it gives people the incentive to start and stop certain AI projects. It also gives the mandate to your community to pause, ideate, and potentially innovate in this space.

7. **AI Usage and Impact Dashboard**

   Develop a metrics dashboard to track AI adoption and outcomes across the institution. Build tolerance to longer-term ROI, as not everything will start yielding immediately. But that will increase your possibility of succeeding if you keep your impact and purpose front and center.

Each element of this toolkit can reinforce the connective capacity at your organization and help move AI adoption from a project or a task to part of the DNA of your organization.

Sidebar: AI in Health and Student Services, A Cautionary Tale

AI's connective reach often extends to some of the most sensitive areas of nonprofit or campus operations: mental health and counseling. Consider medical services and student-support interventions. These domains illustrate both the promise and peril of AI. On one hand, AI can help fill resource gaps or help with predictive analytics to see trends and trajectories that humans might not easily see or account for. On the other hand, mistakes or misuse here carry real potential harm for people's well-being.

Take the case of mental health chatbots that are in growing demand due to the shortage of mental health professionals. The appeal is obvious: The mental health AI chatbots raise huge red flags. A recent study by Brown University in 2025 found that "current AI chatbots routinely violate core mental health ethics standards."[16] In some instances, the AI bots gave advice that reinforced negative self-talk, failed to properly handle crisis situations, or created a deceptive illusion of empathy. Unlike licensed human therapists who are accountable to boards, these AI bots have no equivalent oversight.

Early alert systems are extremely important, thoughtful, and necessary to help humans see patterns, ask a question, or schedule a meeting. They use AI and implement data-driven approaches by flagging "at-risk" students academically or for well-being using predictive analytics or past datasets of former students.

AI is rapidly becoming the connective tissue of mission-driven institutions. The trajectory is clear: In the coming years, virtually in every domain in every nonprofit organization or university, AI will be augmented one way or another. Even as the hype settles on some AI tools and types of AI, the trajectory will keep moving toward more adoption, more deployment, and more training.

Chapter Seven Recap

Core Message

AI can't stay trapped in departmental pilots or one-off tools. To create real mission impact, universities and nonprofits must treat AI as shared institutional infrastructure with common platforms, data standards, governance, and culture so that admissions, student success, health, advancement, and finance all plug into the same safe, coordinated ecosystem instead of a noisy maze of disconnected apps.

Key Takeaways

- Fragmentation is a real risk, not just an annoyance.
- Some institutions, like Vanderbilt, are creating their own product that works for them.
- Agentic AI and giving AI autonomy raises the stakes for governance and culture.

Why It Matters

When AI is treated as institutional connective tissue rather than a pile of apps, organizations can scale experiments without scaling chaos. Fragmented pilots turn into coherent architecture; ethics and governance move from theory into the platform itself; and faculty, staff, donors, and students experience one mission-aligned institution, not a patchwork of tools. Integration would beat isolation, and it's the key path to campus-wide AI that is powerful, safe, and worthy of trust.

Chapter Endnotes

1. Jaime Hunt, "Stop Dabbling with AI," Inside Higher Ed, October 7, 2025, https://www.insidehighered.com/opinion/columns/call-action/2025/10/07/why-higher-ed-must-be-intentional-ai.
2. "Open Enterprise Generative AI," Amplify GenAI, accessed March 1, 2026, https://www.amplifygenai.org/.
3. "University Governance for IT," Clark University Information Technology Services, accessed March 1, 2026, https://www.clarku.edu/information-technology-services/learn-about-its/university-governance-for-it/.
4. Satsawat Natakarnkitkul, "Revolutionizing Higher Education: How Amazon Bedrock AgentCore Enables Production-Ready AI at Scale," AWS Public Sector Blog, August 12, 2025, https://aws.amazon.com/blogs/publicsector/revolutionizing-higher-education-how-amazon-bedrock-agentcore-enables-production-ready-ai-at-scale/.
5. "The Hidden Cost of Disconnected Systems," Civitas Learning, accessed March 1, 2026, https://www.civitaslearning.com/blog/the-hidden-cost-of-disconnected-systems/.
6. Don Snyder (associate dean for development at the College of Connected Computing, Vanderbilt University), interview by Joe Manok and David Woodruff, 2025.
7. "Amplify, Vanderbilt's New Custom Generative AI Software, Is Now Open to All Faculty, Staff and

Students," Vanderbilt University, November 14, 2024, https://news.vanderbilt.edu/2024/11/14/amplify-vanderbilts-new-custom-generative-ai-software-is-now-open-to-all-faculty-staff-and-students/.

8. "Amplify," Vanderbilt University, Amplify GenAI Innovation Center, accessed March 1, 2026, https://www.vanderbilt.edu/agi/platforms/amplify/.
9. Satsawat Natakarnkitkul, "Revolutionizing Higher Education: How Amazon Bedrock AgentCore Enables Production-Ready AI at Scale," AWS Public Sector Blog, August 12, 2025, https://aws.amazon.com/blogs/publicsector/revolutionizing-higher-education-how-amazon-bedrock-agentcore-enables-production-ready-ai-at-scale/.
10. "AWS Generative AI Best Practices Framework v2," AWS, accessed March 1, 2026, https://docs.aws.amazon.com/audit-manager/latest/userguide/aws-generative-ai-best-practices.html.
11. AWS Public Sector Blog Team, "Highlights from AWS Nonprofit Generative AI Week 2025," AWS, October 24, 2025, https://aws.amazon.com/blogs/publicsector/highlights-from-aws-nonprofit-generative-ai-week-2025/.
12. Patrick Frontiera (US higher education strategy lead, Amazon Web Services), interview by Joe Manok and David Woodruff, 2025.
13. Brian Baute, Danielle Rowdy, and Patrick Frontiera, "Doing More with Less in Higher Education: How Institutions Drive Efficiency with AI and Automation on AWS," AWS Public Sector Blog, May 12, 2025,

https://aws.amazon.com/blogs/publicsector/doing-more-with-less-in-higher-education-how-institutions-drive-efficiency-with-ai-and-automation-on-aws/.

14. Maysam Ali (global education GTM lead, Amazon Web Services), interview by Joe Manok and David Woodruff, 2025.
15. "NASFAA 2025 Administrative Burden Survey Reveals Growing Strain on Financial Aid Offices," NASFAA, July 18, 2025, https://www.nasfaa.org/nasfaa_2025_administrative_burden_survey_reveals_growing_strain_on_financial_aid_offices.
16. "New study: AI Chatbots Systematically Violate Mental Health Ethics Standards," October 21, 2025, https://www.brown.edu/news/2025-10-21/ai-mental-health-ethics.

CHAPTER EIGHT

Sector-Wide Shifts and Risks

The sector and technology are moving faster than institutions are. The pace of change is unprecedented, and AI-infused transformation is here. Not only is technological change outpacing organizational change, but our business models and why people give are also changing. There is a crisis in generosity now, and it is telling us that fewer people are interested in giving or volunteering simply because they have an affiliation with an organization or an entity.[1]

Nonprofits are facing mounting pressure to modernize, and donors are expecting more personalization and efficiency. Boards are either tired of hearing about AI or hearing about its successes elsewhere and want AI tools implemented now, while tech companies promise game-changing tools. In practice, doing nothing may pose a greater long-term threat than experimenting responsibly.

One recent case study by the Council for Advancement and Support of Education (CASE) titled "CASE Insights on Artificial Intelligence in Advancement" found that while most advancement professionals are personally eager to adopt AI, their teams and leaders often take a laissez-faire or hesitant approach.[2] Senior leadership is usually not outright prohibiting AI exploration, but neither are they actively promoting innovation, resulting in stagnation by default.

This gap between the sector's fast-changing landscape and institutions' slow internal response is the central theme of this chapter.

Inaction brings its own risks to relevance, effectiveness, and trust. We will examine how the philanthropic sector is shifting around us, what risks and standards are emerging at a sector-wide level, and why thoughtful action is better than complacency. We will argue that embracing AI responsibly with ethical guardrails offers a path to strengthen our missions, whereas avoiding AI entirely can allow external pressures to dictate our fate.

SIGNALS ACROSS THE PHILANTHROPIC SECTOR

Across the philanthropic and higher education advancement sector, clear signals of an AI-driven terrain are appearing. Over the past two years, professional associations and institutions have sounded the alarm that AI is here to stay in fundraising and alumni engagement. At least five national conferences for nonprofits or higher education institutions had "AI, AI, AI" or more in their call for proposals. One had nine AI speakers at one of their events.

The CASE report data reveals a telling pattern: In 82.9 percent of advancement teams, AI use is being driven bottom-up by individual staff experimenters rather than top-down by strategy. In fact, the largest share of surveyed professionals (39.3 percent) said no one in their organization was explicitly championing AI adoption. Only 4.7 percent reported that their team has a formal AI adoption initiative, and merely 13.3 percent have received any formal AI training. The numbers paint a picture of an industry largely in an exploratory phase that is interested in AI's potential but lacking frameworks, training, and executive mandates.

Over the last 12 months, this picture has started to shift. Individuals, organizations, vendors, and engagement platforms are building AI features into their products, nudging nonprofits to try them out. Cara Giacomini, CASE's vice president of data, research, and technology, led a report on generative AI in advancement that was released in the summer of 2025, at the CASE Summit for Leaders in Advancement.[3] The report summarized current use and outlined a roadmap for the next steps for accelerating adoption. This indicates a growing consensus that slow and ad hoc experimentation is not enough, and the sector needs coordinated steps to scale AI responsibly.

External pressures are also signaling change. Donor expectations are evolving; individuals accustomed to personalizing AI-powered experiences in e-commerce or banking will eventually expect similarly savvy engagement from nonprofits and universities. On the regulatory front, data privacy laws and emerging AI regulations are beginning to encroach on nonprofit operations. And peer examples create implicit competition: When one university's advancement office launches a headline-grabbing AI-driven campaign, others feel pressure not to appear antiquated. Simply said, the terrain is shifting, and AI is moving from novelty to norm in philanthropy. But many institutions are still surveying the landscape from the sidelines. They are unsure of best practices, cautious about risks, or are constrained by a lack of skills.

The CASE survey also indicated that while most practitioners are cautious about AI, they are not opposed to it, with only 10 percent having been explicitly advised not to use AI. Instead, there is a vacuum of clear direction. Leaders are largely neither forbidding nor actively encouraging AI, leading to a passive "wait and see" stance. The risk is that by waiting too long, institutions may wake up to find the sector's norms and donor expectations have moved on without them. As Giacomini observed, many advancement leaders won't overcome their concerns until they feel that failure to use AI would leave them at a disadvantage, yet as of early 2024, only about a third of institutions using AI believed they were lagging behind their peers.

CASE and the Push for Sector Standards

Amid this changing landscape, sector leaders are recognizing that individual institutions shouldn't navigate AI in isolation. There is a growing push for sector-wide standards to ensure a baseline of ethical and effective AI use in advancement. CASE, as a leading professional association in educational philanthropy, is at the forefront of this effort. CASE has long played the role of standard-setter. For example, it publishes the CASE Global Reporting Standards that define how fundraising totals are counted and reported across institutions. Those standards are updated every few years, and even incorporate an ethical code of conduct for advancement professionals, underscoring that common principles and definitions are vital in this field.[4] Now with technology's rise, CASE is extending this ethos of shared guidance to the AI realm. Their insights on generative AI reports resulted in guidance on how to adopt AI responsibly.

This included recommendations such as establishing clear data usage policies, creating internal review processes of AI projects, investing in staff training, and starting with pilots that have defined successful metrics.

CASE is also leveraging its volunteer leadership structures to embed AI governance into the profession's DNA. For instance, CASE's Commission on Philanthropy and CASE's Industry Advisory Council have had many conversations on the ethical use of AI. Most recently, CASE announced a new Commission on Advancement Operations and Insights, led by Matthew Eynon, vice president for university advancement at American University in Washington, DC. As part of its charter, the commission is tasked with the fast emergence and growth of AI as a significant opportunity while determining how and where to use AI.[5]

The push for sector standards is also driven by the recognition that AI governance is a team sport. No single nonprofit or university has all the answers to AI's challenges. Through commissions, conferences, and reports, the sector can develop playbooks faster. By aligning baseline standards, institutions can more confidently innovate with AI, knowing that they are not out on a limb alone. They can also help present a united front to stakeholders (donors, regulators, the public), demonstrating that the advancement profession takes responsible AI seriously.

The Ethics of Inaction

As we evaluate the ethics of using AI and how to responsibly use it, we must also evaluate the ethics of inaction.

At its face value, it might be tempting to assume that the safest ethical choice is to do nothing. In this case, and with this technology, inaction has its own perils. In earlier chapters, we introduced the idea that "ethics of inaction deserve the same scrutiny as the ethics of action."

If an AI tool could help us do our jobs better, faster, more insightfully, cheaper, and more personally, and we decide to ignore that potential, are we truly being good stewards of donor funds and the public trust?

Consider a fundraising team that is still using typewriters and only leaving voicemails for donors on their landlines: They wouldn't be able to communicate in a synchronous way with donors on their smartphones, getting responses in minutes instead of weeks. They wouldn't be able to review an email or a correspondence in minutes, instead having to dig through archives to review the original file. When this happened in the 1990s, many resisted it, but at this point, it is considered a highly inefficient way to correspond with donors. Soon, AI will start to feel like a necessity, and those not using it are inefficiently using the donor money to make a limited gift impact. Some institutions, like Clark University, are using AI to increase efficiency, so that frontline fundraisers can send hundreds of handwritten notes to donors. Technology deployment does not mean caring less or automating your relationship, but rather freeing time so you can focus on what matters most: your mission, your community, your donors, and your impact.

The ethical lens demands that we also ask: Are we being fair to our beneficiaries and supporters if we don't leverage available innovations to maximize impact safely and thoughtfully? The moral cost of stagnation is almost overlooked. The philanthropic landscape becomes more competitive; doing nothing could also mean falling behind other organizations that achieve better outcomes for society by using AI, essentially an opportunity cost in impact.

FIVE CATEGORIES OF SECTOR-LEVEL RISK

When considering AI adoption across an entire sector, like the nonprofit or higher-ed advancement sector, it is helpful to categorize the risks at a macro level. Below is a sector-level AI risk taxonomy, grouping the key types of risks that mission-driven institutions face as they integrate AI:

Sector-Level AI Risk Taxonomy

Let us briefly unpack each of these categories with sector-wide examples:

- **Reputational Risk:** This is arguably the most visible in the age of social media and instant communication. A single AI-related lapse at an institution can become a headline that affects public sentiment toward all nonprofits and universities. In an earlier chapter, we shared an example of this misstep when university officials had to apologize for using ChatGPT to craft a message after a shooting on campus. The public and students perceived the message as insincere. For the sector, reputational risk means we must be vigilant that AI never undermines the humanity of our engagement. Building and protecting trust is paramount.
- **Operational Risk:** While much attention is paid to big ethical questions, day-to-day operational reliability is also at stake. Imagine a scenario where a major donor portal is augmented with an AI chatbot to answer questions. If that chatbot malfunctions and gives incorrect information about tax receipts or fund usage without human supervision, it creates confusion and extra work for humans to remedy. Or consider the risk of inaction: An organization that ignores AI might struggle to keep up with the volume of communications or data analysis needed in modern advancement, eventually leading to operational strain, staff burnout, and missed opportunities. Sector-wide, if many organizations lag, the whole nonprofit field starts to look antiquated compared to other industries. On the other hand, racing ahead without proper testing (the chaotic implementation risk) can cause system outages or errors that disrupt services to donors or beneficiaries.

- **Ethical Risk:** This category is broad and underpins several others. Ethical risks include AI perpetuating bias or discrimination (a certain AI model that systematically underrates alumni from a certain group or deprioritizes them for outreach). Privacy breaches are another ethical issue such as exposing sensitive information that violates ethical commitments to constituents. Transparency is part of this, disclosing the use and telling that story to your constituents. Ethical risk triggers serious reputational risk.
- **Compliance or Legal Risk:** The regulatory environment for AI and data is evolving quickly. We already discussed GDPR and California's privacy laws as examples in earlier chapters. If an AI system uses personal data without a proper basis or if data is transferred to an AI vendor in a way that violates a privacy agreement, an institution could face legal action. Additionally, sector-specific regulations can come into play. For example, health-related nonprofits dealing with patient data have HIPAA rules, which restrict how AI can use patient information. We also anticipate future AI-specific regulations, such as transparency requirements or restrictions on certain high-risk AI applications, or having mandatory human review in the process of generating, modeling, or using AI tools.
- **Strategic Risk:** This is the most forward-looking category. It is about the big picture and what matters most. Will the use of AI help us fulfill our mission? An example of strategic risk is misalignment. Implementing AI solely because of its current "hype" or because it's "trendy" could lead to wasted investments or even a mission drift.

By categorizing risks in this way, sector leaders and teams can ensure they're not myopically focusing on one type of risk (say ethics) while ignoring others. A comprehensive risk management approach will ask: How are we safeguarding our reputation and ethical principles? Our day-to-day operations? Are we compliant with laws, and are we positioning ourselves for the future?

The encouraging news is that all these risks are manageable with the right design and leadership. Identifying them is the first step, the next is building the structures, and after that, building the cultures to mitigate them.

Toolkit: Readiness Signals for Chief Advancement Officers, Presidents, and Boards

The following is a toolkit of signals and questions for organizational leaders to assess where they stand and where to focus. You can use these prompts as a checklist in leadership meetings or strategy sessions:

Board Engagement:

- Signal: The board has explicitly discussed AI opportunities and risks within the last year.
- Questions: If AI hasn't made it onto the board's agenda, it's a sign that the institution isn't treating AI as a strategic issue. Boards should inquire: Do we have an AI strategy or policy in place? What is our exposure to AI right now? A proactive board might even form a board subcommittee on technology or require periodic AI risk reports.

Executive Ownership

- Signal: A specific senior leader or cross-functional committee is charged with overseeing AI initiatives and governance.

- Questions: Can the president or chief advancement officer name who is responsible for the AI strategy? If it's diffused or "no one," that implies low readiness. Look for signs like the appointment of an "AI lead" or the establishment of an AI task force.

Policy Framework:

- Signal: The organization has at least a preliminary AI policy or set of guidelines or guiding principles in place, even if this is done informally.
- Questions: Is there a written document or an agreed-upon set of guidelines about how staff should (or shouldn't) use AI tools?

Staff Activity (visibility):

- Signal: Staff use of AI is known to leaders.
- Questions: Can you list some AI tools or use cases people on your team have tried?

Training and Culture:

- Signal: There is an ongoing effort to train staff on AI-related skills.
- Questions: Did the nonprofit send its staff to AI conferences or plan its own training? Did they provide resources for learning (like CASE or AFP webinars or online courses)? Do the team meetings occasionally feature demos of new tech? Is there a climate of fear and skepticism?

Ethics and Risk Protocols:

- Signal: Before launching any AI-related project, there is a process, whether formal or informal, to review risks involving multiple stakeholders (e.g., IT, advancement, legal).

- Questions: What steps do you take to evaluate an AI tool before using it?

Technology Infrastructure:

- Signal: The IT team has integrated or has planned for AI capabilities in the existing systems (CRM, databases, etc.) and has measures for data security in those contexts.
- Questions: Does your data environment support AI, and is your CRM offering AI capabilities?

Early Wins and Learnings:

- Signal: Your organization can point to at least one or two pilots that have been completed and documented, whether successfully or not.
- Questions: What have we learned from our AI experiments so far?

Stakeholder Sentiment:

- Signal: Donors and alumni have shared their thoughts about the use of AI.
- Questions: Have you surveyed your board or donors on the use of AI? Did you have a conversation with your community in a town hall?

Expert Voice: Taylor Swaak

A National Shift in AI Governance[6]

Taylor Swaak's *Adapting to AI* report for the Chronicle of Higher Education offers insights on how universities are translating AI ethics into shared literacy and policy. The report's core message is that most university leaders do not need to become AI experts, but they have a responsibility in learning how to embed clarity, accountability, oversight, and institutional control, not fear. Some of the thoughtful strategies observed when the report was published in June 2024 included:

- Having institution-wide guidelines that clarify how and when to use AI and when it can not be used.
- Committees that are permanent and continuously provide oversight and evaluate risk.
- Literacy programs for faculty and staff that move into cultural confidence and not just compliance.

The report highlights how literacy in AI governance and ethical use of AI begins with institution-wide literacy, starting from the top.

One example that the report cited was the Drexel University permanent watchdog model, a long-standing committee that includes different stakeholders from across the university and provides ongoing oversight on the use of AI in teaching, admissions, research, and meets regularly to evaluate tools and proposed policy changes. Such a model does not slow innovation; it gives it legitimacy.[7]

That sentence could easily appear in any nonprofit advancement shop grappling with AI's emergence. It echoes this chapter's argument that governance isn't punitive; it is enabling. A shared visible structure for AI ethics empowers experimentation, builds trust, and prevents innovation from being either chaotic or exclusive.

The Swaak findings also reinforce one of our chapter's most practical insights: that ethical AI is often less about banning tools and more about scaffolding learning. Colleges that frame AI guidelines as living documents and that train their communities to critically assess AI output are finding fewer compliance battlers and more thoughtful adoption curves.

The future will belong to those who move with deliberate speed and not those who freeze in the past, where they get too comfortable with their operations. Fearless organizations are more likely to risk their reputation, while staying still might risk irrelevance. Embracing responsibility over fearlessness positions the sector to shape its own destiny.

Chapter Eight Recap

Core Message

As our sector shifts faster than our institutions can change to keep up, we have to evaluate the risk of using a fast moving technology like AI, and equally assess the risk of inaction and what that means ethically to our donors and our community when we are not doing things better, faster, smarter, and at a larger and more efficient scale while staying institutional and donor centric.

Key Takeaways

- The sector, technology, donor expectations, and team needs are changing with or without us, and at a pace faster than we have ever witnessed. Doing nothing can erode competitiveness and trust.
- CASE's landmark survey shows that most AI use in our sector is bottom-up pilots, with leadership at most institutions in the wait-and-watch mode.

- Sector standards, like CASE Insights reports, are helping guide the sector.
- Avoiding AI could be as ethically problematic as not using it.
- Risks in our sector include reputational, operational, ethical, legal, and strategic.

In Practice

Ignoring AI is itself a strategic and ethical choice with real costs. The institutions that thrive will combine responsible guardrails, sector-aligned standards, and cultural readiness. The chapter's through-line: The future isn't shaped by the fearless or the ones frozen in the past, but by those who act with deliberate responsibility.

Chapter Endnotes

1. The Generosity Commission, *Everyday Actions, Extraordinary Potential: The Power of Giving and Volunteering* (The Giving Institute, 2024), https://www.thegenerositycommission.org/wp-content/uploads/2024/09/DIGITAL_TGC_FullReport_092424.pdf.
2. Cara Giacomini, Deborah Trumble, Divya Krishnaswamy, and Jacqueline King, *CASE Insights on Artificial Intelligence in Advancement* (Council for Advancement and Support of Education, 2024), https://www.case.org/system/files/media/inline/CASE%20Insights%20AI%20in%20Advancement%20with%20GiveCampus%20March%202024.pdf.

3. Cara Giacomini (vice president of data, research, and technology, CASE), interview by Joe Manok and David Woodruff, 2025.
4. Sue Cunningham, "Building and Protecting Institutional Reputation: What Governing Boards Should Know About Global Reporting Standards for Fundraising," *Trusteeship* 30, no. 6 (2022): https://agb.org/trusteeship-article/building-and-protecting-institutional-reputation/.
5. "New CASE Commission on Advancement Operations and Insights," Council for Advancement and Support of Education, November 19, 2025, https://www.case.org/resources/new-case-commission-advancement-operations-and-insights.
6. Taylor Swaak, *Adapting to AI: How to Understand, Prepare for, and Innovate in a Changing Landscape* (Chronicle of Higher Education Inc., 2024), https://store.chronicle.com/products/adapting-to-ai.
7. "About Artificial Intelligence at Drexel University," Drexel University, accessed March 1, 2026, https://drexel.edu/provost/ai/about/.

CHAPTER NINE

Measuring Impact and Readiness

Most nonprofits are tracking more numbers than ever before with dashboards, KPIs, open rates, predictive scores, and conversion charts. Yet despite all that visibility, many of us still struggle to answer the most basic questions: Is our work improving? Are we more effective? Are we ready for what we've built?

This chapter begins with a simple truth: Readiness and impact are not the same. An organization can be busy, data-rich, and tool-heavy yet unprepared to deliver results that matter most.[1] AI only sharpens, divides, and accelerates activity in ways that can look impressive from a distance but reveal very little progress, trust, or mission outcomes up close. The goal is not to celebrate the dashboards but to interrogate them. It is to replace volume with clarity and mission-driven metrics.

THE MEASUREMENT GAP: FROM ACTIVITY TO INSIGHT

Many organizations are drowning in data but thirsty for insight. It is common to see fundraising and programming teams tracking dozens of indicators only to find that these numbers don't translate into better decisions or mission outcomes. Nonprofit leaders often say we are not learning much from what we measure, and yet few seem to crack the code. Closing the gap requires a cultural shift in how nonprofits approach data: from using it as a reporting tool for boards, leadership, or funders to treating measurement as a learning tool.[2] With metrics, you can ask yourself questions like: Why are we seeing this? What is driving or restraining success? Are we heading in the right direction?

When you use the Fore Field Analysis by Kurt Lewin, you get a more pragmatic view of why metrics often fail to translate into organizational learning.[3] He argues that "every system is held by a dynamic equilibrium of a combination of driving and restraining forces." For us in the nonprofit world, our dashboards typically measure visible surfaces from open rates, gifts closed, or meetings reported. Lewin argues that we should move from measuring things as one-and-done to continuously scanning for what forces impede or accelerate our transformation, and that the value to us is in the diagnosis itself, rather than the resulting score. With insights, we focus on why something happened, rather than focusing on the data and only what happened.

Force Field Analysis turns metrics from static indicators into a map of friction and leverage. It helps leaders see that readiness is not a dashboard story, but a balance of forces problem. When measurement becomes a structured way of surfacing those forces, organizations finally begin to learn and change.

For instance, rather than celebrating an uptick in predictive scores assigned to donors, a team might ask: Did those scores help us steward donors more effectively and ethically? Analyzing the driving forces might reveal that the uptick could have been multifold if a restraining force had been addressed. An AI-ready nonprofit is not just the one that deploys the best AI tools but also governs progress through thoughtfully curated metrics.

So ask yourself, are you focusing on easy wins or sustainable long-term impact? To answer this question, we introduce the AI Risk and Readiness Matrix, a simple two-by-two model that will help you locate your organization's current state and navigate towards "future-ready" impact.

This matrix does not introduce a new tool or framework, but rather links two established frameworks: how well your organization can use AI and how well it can handle the effects of AI. Doing so allows you to visualize where you are and design where you want to go next. When used over time, it can show you a pattern of where you were and where you moved, with either more readiness or more governance. With this matrix, you can start to get a better idea of how progress, risk, and trust are really changing in your organization over time.

INTRODUCING THE AI RISK AND READINESS MATRIX[4]

In earlier chapters, we discussed the AI Readiness State (AIRS) and the Ethics Maturity Spectrum. This framework works by plotting the earlier two frameworks to visualize where you are and what it means to be in one of four quadrants (depending on your score).

On one axis, we have AI Readiness from low to high: how far along your organization is when it comes to integrating data, building skills, or adopting tools. On the other axis, we have the Ethics Maturity (low to high), which encompasses ethics, oversight, and alignment with values. In other words, how mature and structured you approach managing AI's risk and responsibilities.

Advancement AI Risk and Readiness Matrix

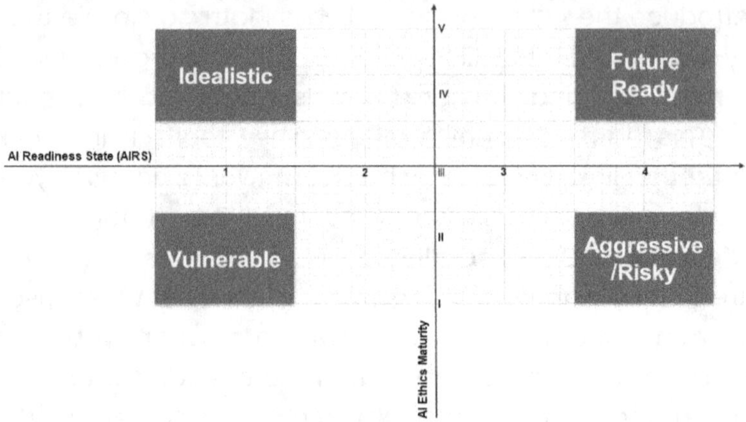

By plotting these two dimensions, we reveal the four quadrants that describe an organization's overall posture:

- Vulnerable (Low Readiness, Low Governance)
- Aggressive/Risky (High Readiness, Low Governance)
- Idealistic (Low Readiness, High Governance)
- Future-Ready (High Readiness, High Governance)

Before diving into each quadrant, it's worth noting that these aren't fixed labels or judgments. They are states, a snapshot in time. Organizations can move from one quadrant to another, and the goal of measurement is to help navigate that journey.

Vulnerable Quadrant (Low Readiness, Low Governance)

The Vulnerable quadrant is the most precarious place to be. Organizations are neither prepared for technological change nor particularly ready to manage the ethical challenges that technology brings.

Culturally, a vulnerable organization is often reactive or stretched. They haven't invested much in AI or data infrastructure, not out of deliberate caution, but more often due to lack of capacity, knowledge, or urgency. They are mostly the "we will cross that bridge when we get there" mindset, but without realizing that some bridges may collapse before you get to them.

When vulnerable organizations engage with AI, it is usually not through a strategic pilot they designed, but rather through external pressures or one-off experiments. For instance, a well-meaning board member might donate a subscription to some AI tool and ask them to try it or use it, or a staff member might independently start using ChatGPT to draft grant proposals because it helps them save time or get ideas. These engagements are often uncoordinated and unguided.

Pilot outcomes in this quadrant are therefore a hit or miss and sometimes downright invisible. Nothing major may go wrong immediately, but risk accumulates under the radar. Without governance, there is likely no data consent for complex tools, no bias check on AI-generated content, no cybersecurity review of that free AI app now plugged into their database. The organization might not even be measuring results from these mini experiments. A staffer could be thrilled that an AI writing assistant saved them hours, with a hidden efficiency gain, but leadership wouldn't know because no one thought to capture that metric. The metrics focused on vulnerable organizations tend to be on traditional outputs like dollars raised or people served with little granularity.

Leadership in vulnerable organizations is often in a survival or status quo mindset. They're likely to be passionate about the mission but lack either the knowledge or the bandwidth to pursue digital transformation. Discussions about AI readiness or ethical technology governance might rarely reach the board agenda. It is not that these leaders don't care about ethics or efficiency, but day-to-day is consumed with things they feel are more important.

Aggressive/Risky Quadrant (High Readiness, Low Governance)

Organizations in the Aggressive/Risky quadrant are racing ahead on the AI curve without adequate oversight. They have high technical readiness with plenty of pilots and projects, maybe a data scientist on their team, and leadership that is enthusiastic about innovation. Aggressive/risky teams idolize speed and efficiency with the "ask later" philosophy. There is a natural bias for action in these environments, which can be energizing, but also an aversion to pause for reflection. Risk management is treated as a blocker rather than a prerequisite.

What happens when such a culture runs AI pilots? Pilot outcomes in this quadrant tend to be mixed, technically impressive but fraught with unanticipated issues. On one hand, these organizations might see quick wins, say an AI model that identifies new major gift prospects that the human team overlooked, or a chatbot that suddenly handles 80 percent of donor inquiries, freeing staff time.[5] These results show up as glowing metrics on efficiency, more tasks handled per person, faster response times, or even lower operational costs. But because governance is minimal, downsides emerge just as quickly. The Aggressive/Risky quadrant is likely to result in an ethical lapse of judgment. Perhaps the chatbot shared private sensitive data, or said something inappropriate, or a prospecting tool omitted certain constituencies altogether. Without proper design, strong oversight, and "kill switches," this category is high risk for nonprofits.

Leadership traits in this quadrant are often characterized by visionary zeal without sufficient patience. Leaders here are bold and opportunistic, and see the potential of AI everywhere. They're eager to implement it without counterbalancing that with the cautiousness necessary for responsible use of AI.

Moving an organization in this quadrant to become future-ready is not meant to slow down the organization, but rather to solidify where the organization is and make sure it reduces its risk and exposure for responsible use.

The Idealistic Quadrant (Low Readiness, High Governance)

At the opposite corner of the matrix lies the Idealistic quadrant. These organizations have high ethical governance relative to their peers, but low AI readiness; they are technologically cautious or lagging. They don't have to be against technology; they might be very technologically advanced, but they prioritize what they can't and shouldn't do rather than identifying what can be done in an ethical way. To them, zero risk is best managed by doing nothing.

An Idealistic nonprofit emphasizes values, mission integrity, and thoughtful deliberation in every decision. They likely consider the AI hype as something they have seen before and think that it's either likely to go away or they should wait for others to use it first. When an idealistic organization does run an AI pilot, it tends to be small-scale, overly controlled, and aimed at affirming values more than exploring innovation. It is mostly prejudged through confirmation bias to initiate the pilot, knowing the learning outcomes before even starting.

The positive part is, these pilots are usually very stakeholder-friendly and highly ethical. They are the safest place to start.

Yet, the impact metrics (efficiency gained, funds raised, constituents served better) may be minimal or nonexistent because the technology application was so limited. In other words, idealistic pilots are safe (sometimes so safe that they fail to demonstrate any significant benefit or value). This often creates a self-fulfilling prophecy: The leadership says "AI is not worth it for us" or "this does not work for our environment" because the very constrained experiment unsurprisingly yielded constrained results with very limited or no ROI.

Institutions in this quadrant are usually minimalists and focus only on the mission and keeping it simple. Or they have a moral issue with using AI. For instance, they prioritize the carbon footprint of using AI (even though they travel for work using planes, commute to their work, and use video calls). Some institutions have a backstory, like a recent cyber-attack, or donors were upset about a national issue, or their alumni were upset they took money from the wrong sources. During that repair phase, the ROI from AI is lower to them than the cost of taking any risk.

Leaders in this Idealistic quadrant are principled stewards. They are often long-tenured nonprofit executives or board members deeply rooted in the mission. They take their role as guardians of the organization's soul very seriously, which is commendable, but their approach to it is through being idealistic about technology, rather than using technology to get there. Their comfort zone is crafting policy and not prototyping new technologies. Their decision-making style usually looks at the moral lens without equally engaging with strategy or experimentation.

They likely adopted the internet tools that we use today when almost everyone else did, and when it became more than an issue for survival.

The good news for this quadrant is that it's much easier to build the readiness muscle than the ethical muscle, and opening up to what they are comfortable with, bit by bit, could help, but would require them to be comfortable with the innovation cycle and accept the notion of failing fast and scaling what works.

The Future-Ready Quadrant (Low Readiness, High Governance)

Welcome to the quadrant that we all should aspire to be in: Future-Ready. Here we find nonprofits that have achieved a healthy equilibrium between technology and trust, and readiness and proper governance. Future-ready teams are strategic, learning-oriented, and grounded in the mission. They encourage experimentation with new technologies but do so within an environment of accountability and reflection. Their mindset is neither "move fast and break things" nor "move slowly to avoid all risk," but rather the "move thoughtfully and consistently towards greater impact" mindset.

These organizations treat AI initiatives as a microcosm of their future operating model; they define success before starting a project, and they define "kill switches" or a criterion to pause as a safeguard. By training the team on both the tools and the operating model, we ensure they can innovate and experiment within a sandbox, whether they succeed or fail.

A future-ready team, in this sense, is within an office that benefits from the innovation ecosystem in day-to-day work, contributes to the success of it by sharing feedback, optimizes the process of innovation, and pilots their own ideas so that others can also benefit from their collective innovation. When they leave the team, they will also become agents of change in their new institution so that other nonprofits become future-ready. This cycle of innovation is moving so fast that it will likely take a while before we slow down again. Similar to COVID-normal, this might be the AI-normal.

Leadership in the Future-Ready quadrant prioritizes what we call "ethical innovation" over trends and profit, remaining focused on the impact and mission. These leaders articulate a clear vision of how AI and data serve the mission, and where they do not, and they champion both performance and principle. You will often find that they have multidisciplinary knowledge, perhaps a CEO who understands program work and has taken time to learn about AI capabilities, or a chief information officer (CIO) who works hand-in-hand with a chief ethics officer, or a committee, or a working group on AI governance.

Such leaders foster a culture of continuous readiness and treat AI readiness not as a one-time state or certification, but as an evolving competency they invest in and their staff on an ongoing basis. Be it learning about what is new, what is upcoming, staff training, policy updates, or audit processes. They know the goalposts will keep moving for AI and for our future-ready campaigns and future-ready teams, and they keep adapting, adjusting, and innovating accordingly.

To maintain a future-ready status, these organizations often develop frameworks and operational plans that they keep updating monthly or quarterly. This quadrant is aspirational, but attainable, even for organizations that might have just started.

The GlobalPhilanthropy.ai Readiness and Ethics Assessment Tool

We built GlobalPhilanthropy.ai to offer sector-specific resources and assessment tools designed to help universities and nonprofits evaluate their AI readiness through the dual lens of capability and ethics.[6] Built on the same conceptual architecture used throughout this book, including the AI Readiness State (AIRS) model, the AI Ethics Maturity Spectrum, and the AI Risk and Readiness Matrix. The tool translates abstract concepts into a structured diagnostic of 50 questions that help evaluate an organization's data hygiene, workflow integration, governance structures, risk posture, and cultural readiness. It shares a specific score and proposes immediate next steps that can be deployed over the next three to six months.

For nonprofits navigating early-stage experimentation or accelerating toward scaled adoption, the GlobalPhilanthropy.ai assessment functions as a grounding mechanism. It provides a shared language for boards, executives, and staff, and it allows leaders to benchmark progress, track improvement over time, and ensure that the pursuit of AI-enabled efficiency never outruns the organization's ethical footing. In short, it measures readiness not as a technological milestone, but as a leadership commitment to responsible, mission-aligned innovation.

METRICS THAT MATTER: QUALITY, INCLUSIVITY, EFFICIENCY, REVENUE

Having established that Readiness x Ethics = Impact, we turn to the practical question of: What should we measure to gauge that impact? We all know that not all metrics are created equally, and we will only discuss here the categories that we think matter most. These are quality, inclusivity, efficiency, and revenue.

- **Quality:** This is about answering questions like: Are we doing better because of AI? This might include metrics related to improvement in program outcomes, accuracy of insights, or better donor satisfaction.
- **Inclusivity:** It is not just about what impact we achieve through AI but how we achieve it and for whom. AI systems amplify things, and you want to make sure you are not amplifying bias or exclusionary behavior.
- **Efficiency:** This is where AI can shine the most with its current form. With limited time and financial resources, quantifying this part is perhaps the most visible set of metrics.
- **Revenue:** Finally, metrics related to revenue track how much AI initiatives are directly affecting our financials, from fundraising outcomes to more student yields in admissions.

These four categories create a balanced approach for AI impact, and they interrelate to show you a more holistic view of your AI efforts.

The intangibles are also important to track, from how many people joined your team because it aspires to be future-ready, to staff innovation, well-being, and the ability to pilot outside of pure AI use cases. A lot of AI success depends on scaling pilots and use cases, and you can't scale what you don't measure. You can make sure that you are scaling what should be scaled and what matters most to make a difference in your mission, not just for the sake of having big numbers.

Case Study: The Children's Hospital of Philadelphia (CHOP)[7]

To see how these metrics work in real life, let us look at a case study of CHOP that shows how AI can be used to increase fundraising impact and how to measure the results. CHOP is the nation's first pediatric hospital and a world-renowned institution. Like many nonprofits in healthcare, its mission is to save children's lives, and achieving that mission requires significant philanthropic revenue.

The challenge: CHOP's advancement team wanted to maximize fundraising outcomes (revenue) by better understanding their vast constituent base. Traditional ways of dividing up donors and marketing to them were not very useful because they put donors into broad groups and might have missed what really drives each supporter. CHOP serves diverse supporters of grateful patients, community philanthropists, and grateful parents, and needed a way to authentically connect with each at scale. The hospital also had ambitious goals to accelerate major gifts (efficiency) and ensure no interested supporter "falls through the cracks" (quality). Essentially, they asked: Could AI help us listen to our donors better, personalize engagement, and raise more funds faster?

A solution came for CHOP with their partnership with both Moore (a tech firm) and Microsoft's Azure.[8] This allowed them to process massive amounts of data with machine learning and detect previously hidden patterns in donor behavior.

The results: The impact of CHOP's AI-driven fundraising has been remarkable. By embracing data-driven personalization, CHOP achieved an 85 percent increase in donor response and a 25 percent boost in average gift size. It also helped donors move through the major gifts process 17 percent faster. Some relationships now take weeks instead of months to move through the major gifts pipeline, thanks to timely insights and efficiency gains from the development team. The metrics translate to millions of additional dollars raised. Beyond the numbers, CHOP also saw qualitative improvements (quality) by linking people and causes in a more individualized way, creating more meaningful donor experiences.

Another unique route of AI is one that incorporates QUANT (quantitative computing) to do complex predictive analytics. They have 500 or so variables that are related to the US economy. The healthcare sector (specifically in Philadelphia), philanthropy overall, their team members, and the actual giving pipeline can relate to the market better and predict their financial goals in the development office with much more accuracy than any other institution we researched. This included achieving accuracy of hundreds of millions worth of financial goals with an error margin of $5,000 in the most recent year and a very close result the year before.

Businesses do this all the time, particularly hedge funds, as jokingly shared by Jon Thompson, the associate vice president for philanthropic strategy and technology at CHOP. We watched the Showtime TV show *Billions*, which features the inner workings of a hedge fund, and one of the characters created their own QUANT model. We wondered: Why can't we do this ourselves for a hospital? The data can be fully anonymized, since most of the 500 variables have nothing to do with CHOP or its specific donor base, but instead reflect variables that impact the environment in which the CHOP-specific donor base is operating in a particular year.

Thompson said that the best way to do this scale of AI is by listening to data. Listening is one of the key elements of the soft skills we discuss in the following chapter, and it is what differentiates the CHOP approach from many others when it comes to measuring impact and ROI. The use of high ethical standards along with high readiness enabled CHOP to work in a future-ready quadrant that meets the for-profit corporate standard of future-ready on this front. The listening aspect of their work allowed CHOP to use data to hear constituents' needs better, and not as a gimmick. Another lesson was the new pathways and inclusivity. By moving beyond conventional segmentation, CHOP engaged a more diverse supporter base, identifying those who were "lost" over time or were not even known to them in the first place.

Simply said, CHOP shows that when a nonprofit is AI-ready and measurement-minded, the rewards are significant: improved quality of engagement, greater inclusivity in outreach, major efficiency gains, and substantial revenue growth.

Reflecting across all the chapters, a pattern emerges: Measurement and learning are the golden thread. Whether it was in fundraising, ethics, design, team, campaigns, integration, or risk, every theme required feedback loops and evidence to truly succeed.

This reinforces one of our core threads that an AI-ready nonprofit isn't one that blindly jumps on technology, but one that builds a culture of continuous improvement fueled by data and guided by ethics.

Chapter Nine Recap

Core Message

AI doesn't automatically create impact just because it is present in your technology stack. Real impact happens when readiness (capability) and ethics (governance) come together and when organizations measure what truly matters: quality, inclusivity, efficiency, and revenue, not just activity. In other words, you can't scale what you don't measure, and you can't trust what you don't govern.

Key Takeaways

- Readiness does not equal impact.
- Many nonprofits are "drowning in data but thirsty for insight."
- Force Field Analysis measures forces, not just scores.
- Future-ready organizations experiment with AI inside clear guardrails.
- GlobalPhilanthropy.ai Readiness and Ethics Assessment tools help diagnose, score, and provide specific next steps to move towards a future-ready status of AI and ethics readiness.
- Four categories keep AI impact honest and mission-aligned: quality, inclusivity, efficiency, and revenue.
- CHOP demonstrates what happens when an institution "listens through data."

Chapter Endnotes

1. Nathan Furr and Andrew Shipilov, "Beware the AI Experimentation Trap," *Harvard Business Review*, August 29, 2025, https://hbr.org/2025/08/beware-the-ai-experimentation-trap.
2. Michael Wilkinson, "From Vanity Metrics to Mission Metrics, Rethinking Nonprofit Data," Torchbox Charity, May 29, 2025, https://torchbox.com/charity/blog/from-vanity-metrics-to-mission-metrics-rethinking-nonprofit-data/.
3. "Force Field Analysis," Institute for Manufacturing, University of Cambridge, accessed February 28, 2026, https://www.ifm.eng.cam.ac.uk/research/dstools/force-field-analysis/.
4. Joe Manok, "What Is Your Next AI Move?" *Forbes*, September 29, 2025, https://www.forbes.com/councils/forbestechcouncil/2025/09/29/what-is-your-next-ai-move/.
5. "Gartner Predicts Agentic AI Will Autonomously Resolve 80% of Common Customer Service Issues Without Human Intervention by 2029," Gartner, March 5, 2025, https://www.gartner.com/en/newsroom/press-releases/2025-03-05-gartner-predicts-agentic-ai-will-autonomously-resolve-80-percent-of-common-customer-service-issues-without-human-intervention-by-20290.
6. "AI Assessment Tool," GlobalPhilanthropy.ai, accessed February 28, 2026, https://www.globalphilanthropy.ai/assessment-tool1.

7. Jon Thompson (associate vice president for philanthropic strategy and technology, Children's Hospital of Philadelphia), interview by Joe Manok and David Woodruff, October 2024.
8. "Children's Hospital of Philadelphia Enhances Fundraising Efforts with Moore, a Microsoft partner, and Microsoft Azure," Microsoft, July 15, 2024, https://www.microsoft.com/en/customers/story/1787571963406239797-chop-azure-nonprofit-en-united-states.

CHAPTER TEN

The Skills to Outlast Any Algorithm

Even as artificial intelligence accelerates insight, scale, and precision in fundraising, the long arc of philanthropy remains fundamentally human. In donor meetings, boardrooms, and community forums, what ultimately moves people is not an algorithm's output but a sense of trust, empathy, and understanding. At the outset of this book, we asserted that while embracing AI is imperative for modern nonprofits, humanity is the strategic differentiator and moral counterweight. The next decade of philanthropic leaders will be shaped not by who automates the most, but by who listens the best. The most enduring strengths of a nonprofit, such as trust, empathy, relationships, curiosity, humility, values, purpose, or legacy, cannot be automated. These skills do not appear on a technology roadmap, and yet they determine whether sophisticated analytics translate into sustained support and better impact. AI-driven tools can analyze donor data and personalize outreach at a previously unimaginable scale, and these technologies are just the start.

Leaders in the social sector face a paradox of progress. Qualities like generous listening and human presence are not antiquated, or "nice to have" in the AI era; they are the essence of what matters most. They can be amplified or accelerated through AI-driven insights or tools, and remind us that real human engagement with donors is the real end.

RECLAIMING ATTENTION IN AN AGE OF ACCELERATION

One of the defining challenges for nonprofits today is reclaiming genuine attention in an age defined by acceleration. We live in a time when multitasking is a badge of honor and distractions are relentless. This tension between efficiency and humanity must be managed carefully. Digital overload is the norm, and AI is only going to expand on that, making genuine interactions increasingly scarce and, therefore, an extraordinarily powerful tool.

Consider the following donor experience: If they receive communications that are perfectly timed and highly personalized, this might seem interesting once or twice, but for it to remain meaningful and grow the relationship itself, it has to be augmented with a real human connection. Now consider the opposite, where an AI model helps identify a potential major donor, after which a frontline fundraiser, using other AI tools, curates personalized outreach and builds a deeper relationship with that person and listens to them, going beyond the data they are generating. That meaningful attention builds the relationship.

Reclaiming attention also has an inward-facing dimension for leaders. Nonprofit executives often feel pressure to move faster, implement the next tool, respond to the latest trend, and process the never-ending influx of information. With that, we still need moments of reflection for effective leadership. When we reclaim attention, we resist the default mode of perpetual partial attention and instead model a presence that others in the organization can emulate. This kind of presence is contagious.

Part of reclaiming attention is about pausing and asking. No one illustrates this better than Hal Gregersen of MIT Sloan School of Management, author of *Questions*

Are the Answer.[1] Gregersen has spent decades researching the power of questions in leadership. The courage to ask questions, particularly fundamental and uncomfortable ones, is a hallmark of transformative leaders. He makes a compelling case that curiosity is a discipline that leaders can cultivate, starting with the willingness to say "I don't know" and listen genuinely to others. In the context of a nonprofit, a leader might pose a catalytic question like: What are we missing in how we engage our donors? The courage to ask also entails listening to answers that may be unexpected or challenging. In the era of prompting, we are witnessing firsthand how questions (directed to these AI tools) are the only pathway to the answers that they provide. AI and humans can provide many answers, but leaders must supply the questions.

GENEROUS LISTENING: SELF, OTHERS, WORLD[2]

What does it really mean to listen generously and in a comprehensive way? The term implies giving: giving a person's time, attention, and compassion to the act of listening. Generous listening involves behaviors like being fully present, eliminating distractions, suspending judgment, approaching conversations with an open mindset, and listening beyond words to the emotions, tone, and body language.

Turkish artist and philanthropist Vuslat Doğan suggests that generous listening has three dimensions: listening to self, to others, and to the world. This framework is a helpful guide for nonprofit leaders striving to maintain human-centered awareness alongside AI-augmented workflows. Each level of listening strengthens a different facet of leadership, and together, they form a kind of "listening intelligence" that no algorithm can replicate.[3]

In an interview, Doğan pointed out that many of today's societal and workplace challenges, be it disengagement or social polarization, stem from a failure to truly listen. Leaders often miss signals, whether of discontent or creativity, simply because they haven't cultivated a culture of attentive listening. Doğan notes, for instance, that phenomena like the "Great Resignation" and "Quiet Quitting" were, in part, a wake-up call that employees did not feel heard or valued in their organizations. No amount of AI or analytics on employee productivity will reveal the problem if leadership isn't listening to what their people need.

Listening to Self

This is the foundational level of listening and involves being attuned to your inner voice, values, and biases. This requires you to carve out real space for honest reflection, to ask yourselves why you made a certain decision, whether the decision aligned with your ethical principles, where you feel defensive, and what that might be telling you about yourself or the situation.

Self-listening requires humility, the willingness to acknowledge what you don't know, and the courage to ask for help or input. It is also a practical safeguard in the AI era. When you are grounded with your values, you are less likely to blindly follow algorithms. Listening to yourselves creates focus on mission and impact, rather than on dashboards and quarterly results.

Listening to Others

Listening to others is the form of listening we are most traditionally used to, with focus on our community of colleagues, volunteers, board members, and partners.

Doing that in a generous way involves being fully attentive to the other person's words, needs, and emotions. It could involve asking a donor why they give to your organization, holding open forums for colleagues to share their hopes and worries about new AI tools, or asking frontline staff if they have seen something that you may have missed, and then pausing and not arguing with the answer. This form of listening creates both psychological safety and relational intelligence. Technology can analyze thousands of surveys in seconds, but it cannot replace the silence as someone collects their thoughts or the eye contact between two people.

In fundraising, this is where "people give to people" becomes concrete. AI might identify who is likely to give, but listening to others is how we learn why they might give, and how we honor that motivation over time.

Listening to the World

This is where we listen to our overall environment and look at the bigger picture from global trends, listening to nature, demographic shifts, public perceptions, or policy changes.

In an AI-enhanced organization, the temptation is to equate "listening to the world" with monitoring feeds and dashboards, such as news alerts, economic indicators, social media trends, and AI-generated forecasts. Those inputs matter, but generous listening goes a step further: Your organization must learn to interpret what those signals mean for your mission and for the people you serve.

Putting Generous Listening to Work

The three dimensions reinforce one another, with leaders who listen to themselves being more grounded, less defensive, and better able to truly listen to others. Those who listen to others gain empathy and insight, which improves their ability to listen to the world. Those who listen to the world bring back the big picture and context that can reshape their inner assumptions and challenge their own thinking.

In an age where AI can "hear" more than any human by transcribing meetings, searching the web, and summarizing sentiment, the real differentiator is not raw input: It is how we listen and what we choose to do with what we hear.

EQUITY, DIGNITY, AND THE WORK OF ALEX AMOUYEL[4]

Alex Amouyel, the president of the Newman's Own Foundation and formerly the executive director of MIT Solve, has a central theme in her work: Technology and innovation should serve human dignity and equity. Amouyel, author of *The Answer is You*, believes that solving the world's toughest problems requires not just clever ideas or new tech, but a steadfast commitment to values like equity, inclusion, and moral purpose.[5] In other words, impact begins with who we are and what we stand for.

The book's message resonates directly with the concept of personal responsibility and agency in driving change. Rather than looking for some external savior, be it a technology, a mega-donor, or a government intervention, she emphasizes that each of us has a role to play by aligning our skills and experience (our "superpowers") to our actions and values. If equity is a stated value, a leader must listen to whose voices are missing in their decision-making processes and actively bring them to the table. If dignity is a value, leaders must examine whether their use of AI and data treats people as distinct individuals or reduces them to statistics.

One way Amouyel has tied innovation to equity is by advocating inclusive innovation. This means involving those who are most affected by a problem in designing the solution. In practice, for a nonprofit, that could mean co-creating programs with the very community members who will benefit from them, rather than imposing a top-down solution. At minimum, it means listening to community members and recruiting diverse reviewers when evaluating innovations, something often overlooked in the excitement of a new predictive model or a new app.

Dignity in a fundraising context also plays out in how we treat donors and community members as partners rather than targets. Amouyel's stance on moral purpose suggests that fundraising should never lose sight of the real people involved. For instance, just because AI can segment and score donors by their wealth doesn't mean a nonprofit should treat high-net-worth individuals as mere dollar signs. Dignity means recognizing the intrinsic worth of every supporter, big or small, and every person served. It means ethically using AI, ensuring transparency and respect in how we handle data.

Amouyel also highlights the importance of purpose-driven innovation. In the rush to adopt the latest tools, she recommends asking: "To what end?" Consider whether implementing an AI solution will advance the mission in a way that aligns with the core purpose. If a particular innovation doesn't serve equity or dignity, is it really innovation or just novelty? For example, an AI chatbot might make donor communication more efficient, but if it inadvertently makes the experience colder or more exclusionary for some donors (say, those who prefer human contact), then the net effect could be negative.

Leaders have to weigh these considerations, and Amouyel's approach implies that sometimes the "innovative" choice is to improve a human process rather than adopt a high-tech one if that better serves equity and impact.

In her book, Amouyel sees purpose as the intersection of a team's superpowers and a real community need. That idea translates well into organizational leadership of a nonprofit, where the best intersection is that of a nonprofit's capability, its mission, and the needs of the community it serves. Aligning this purpose then becomes the sweet spot for innovation.

"PEOPLE GIVE TO PEOPLE" IN THE AI ERA

The classic fundraising statement, "people give to people," has long conveyed that donations are fundamentally personal and donors give because they feel a connection to the beneficiaries or the values of your nonprofit or its community.

The AI era challenges us to reinterpret "people give to people" as a strategic principle: How do we ensure our increasingly digital fundraising efforts remain authentically human and relational?

Remember that as you use more AI tools, you are not the only organization sending your donors hyper-personalized messages. Almost everyone else is doing so. The key will always be the human touch to make a personalized outreach feel more authentic. Research in philanthropy consistently shows that factors like trust, recognition, and a sense of belonging are key drivers of donor loyalty and generosity.

The AI era does not diminish the truth that philanthropy is personal; it amplifies it.

Chapter Ten Recap

Core Message

As AI expands speed, insight, and personalization, the enduring differentiators in philanthropy remain unmistakably human: listening, attention, presence, values, and meaning. The nonprofits that thrive will be those that use AI to enhance and not replace the deeply human work of trust-building.

Key Themes

- Reclaiming attention in an age of acceleration.
- Generous listening to self, others, and the world.
- The courage to ask which questions are the answers.
- Innovation only matters when it affirms human dignity.

Chapter Endnotes

1. Hal Gregersen, *Questions Are the Answer: A Breakthrough Approach to Your Most Vexing Problems at Work and in Life* (Harper Business, 2018).
2. Vuslat Doğan Sabancı (founder, Vuslat Foundation), interview by Joe Manok and David Woodruff, 2025.
3. Vuslat Doğan Sabancı, "'Generous Listening': The Skill You Need Most in the New World of Work," World Economic Forum, January 6, 2023, https://www.weforum.org/stories/2023/01/generous-listening-work-skills-leadership-davos23/.

4. Alex Amouyel (president and CEO, Newman's Own Foundation), interview by Joe Manok and David Woodruff, 2025.
5. Alex Amouyel, *The Answer Is You: A Guidebook to Creating a Life Full of Impact* (Jim Dandy Publishing, 2022).

THE LAST WORD

AI is no longer a disruption on the horizon; it is here as a permanent condition of our environment. Over the course of the book, we've seen that AI-ready is not about adopting new tech trends; successful AI adoption requires organizational integration. The true challenge is cultivating leadership and culture to wield AI tools wisely without losing the human touch, the backbone of philanthropy.

A FUTURE-BACK VIEW: THE NEXT DECADE OF MISSION-DRIVEN AI

Becoming AI-ready is ultimately about becoming change-ready as an institution. Remember Ron Shaich's future-back approach. As we stand at the edge of this new frontier, it helps to imagine the world in 2030. It is difficult to predict how fast innovation and adoption will take place, as well as what an AI-ready nonprofit will look like then. We work backward to today. This isn't about predicting flying cars or sentient AI fundraisers; it is about anticipating the leadership reflexes and organizational norms that will define success in the next chapter of our work.

So, what will define the AI-ready nonprofit in 2030?

Trust, by far, appears to be the most competitive advantage. Nonprofit organizations that consistently demonstrate transparency, privacy, and respect in their interactions with donors, including how they use AI, will stand out as beacons of credibility. In practice, an AI-ready nonprofit of the future might publish an annual "AI Accountability Report" to its stakeholders, for example, showing how they use AI in service of its mission, much as they issue audited financial statements.

Similar to the growth of internet use over the past decade, AI readiness will soon be part of the broader realm of organizational excellence. AI will become part of our work, just like we do not have a task force for emails or websites, we will no longer need AI task forces. New leadership reflexes will center on agility and reflection. Rather than one big "AI transformation," organizations will normalize continuous evolution. The habit of reflecting and adjusting will become more of a norm than what we have now.

Evolving expectations of communities and donors will shape this future as well. By the next decade, people who engage with nonprofits will expect a higher degree of personalization, interactivity, and responsiveness than ever before, similar to what they are getting and will be getting from consumer applications, products, and services.

Mission-driven organizations will still be standing for the same core values as they do today. This will likely become even more prominent, more important, and more visible. As we close, we ask you to consider this a personal charge from us to you, a fellow leader, practitioner, or changemaker. We have explored concepts, frameworks, and examples, but what matters most are actions. We hope that you start or continue to build in the coming years, beginning with a culture of learning, curiosity, and experimentation for you and your team. Remember to:

- Design with ethics, transparency, and humanity at the core.
- Adopt technology with intention and no impulse.
- Listen generously to those you serve, to your inner voice, and to the world around you.
- Build teams capable of curiosity and change.

Let AI free you to do the work only humans can do. Use automation and intelligence to free your time for the people-to-people relationships that matter most. None of this requires advanced degrees or huge budgets, but a commitment to change and learn.

In the end, the mission-driven organizations that will thrive won't be those with the fanciest algorithms, but those with the wisdom, clarity, and humanity to use them well.

ACKNOWLEDGMENTS

Joe: Thank you to . . .

- Jomana Qaddour, who suggested the concept of the book and inspired me to write it.
- David Fithian, president of Clark University, who started this conversation with a simple and powerful question.
- Clark University, a medium that enabled innovation and the ability to experiment, fail fast, learn, and develop.
- Brent Grinna, founder of EverTrue, for being an engine of innovation for our industry.
- Nour El Hoda Afram, my mother in Lebanon, who always inspired me to be better.

David: Special thanks to . . .

- Adam Martel and Rich Palmer, co-founders of Gravyty, for jumpstarting my interest in AI.
- Denise Woodruff, Natalie Hachem, and John Hachem for supporting my incessant curiosity and providing loving support.
- The numerous mentors in my career, who instilled in me a passion for lifelong learning.

And a very special thanks to Christopher Cannon for his insights and review of the book.

FRAMEWORK APPENDIX

The AI Readiness State (AIRS) Framework

- **AIRS 1: Reactive**
  The organization is skeptical, fearful, or unaware of AI, with little to no use of AI.
- **AIRS 2: Exploring**
  Some pockets of staff or departments may be AI-curious, using tools like ChatGPT for themselves. There is no formalized policy on AI, only cautious experimentation.
- **AIRS 3: Operational**
  Leadership and the institution have recognized AI as a strategic priority and have taken initial coordinated steps, with potential pilots and policies underway.
- **AIRS 4: Strategic**
  AI has become a core driver in the organization, fully aligning with the mission through daily operations, planning, budgeting, and leadership priorities. The goal is to augment human intelligence with machine support to make the institution smarter, more agile, and strategic.

The Narrative Intelligence Loop (Data, Story, Action, Feedback)

The Narrative Intelligence Loop is a type of iterative decision-making cycle adapted to a philanthropic context where story, not data alone, is the engine of action.

- **Data** is where the loop begins, in real signals donors produce every day: quantitative behavior (a gift, an event registration, a pattern of email openings) and qualitative cues (comments in a survey, a theme from a contact report, or a public professional update).
- **Story** is where creativity and strategic intent take over. Data alone doesn't inspire a gift, but it does reveal which needs are urgent and where donor values align with institutional priorities.
- **Action** is the deployment of that story across channels, like email, phone, text, proposals, events, or volunteer outreach. Teams respond to donor behavior as it happens and when it is still fresh in the donor mind.
- **Feedback** closes the loop. Every action produces information: Did the donor attend, respond, give, decline, or offer feedback? Did the story resonate, fall flat, or end up trending? Did a new segment outperform? AI can quickly summarize these patterns, surfacing which messages convert, which need revision, and which donors require personal follow-up. This allows for continuous learning.

The Four Data Frontiers Framework

- **Private Donor Data (Gifts, Notes, Events, Surveys)**
  - The data in the system that relates to an individual's donor engagement. This information can help predict future behavior and provide insights that help the staff act sooner, with more empathy, personalization, and relevance.

- **Public Donor Data (Promotions, Media, Boards, Visible Philanthropy)**
  - Information available in the public domain, such as news articles, social media profiles, corporate websites, philanthropy, and political contributions, that can be used to refine donor engagement.
- **Private Institutional Data (Priorities, Proposals, Research maps)**
  - Internal planning documents, reports, campaign priorities, unpublished proposals, and research initiatives that need funding. This is basically the institution talking to itself about what it could accomplish if resources are allowed and available. This creates a win-win scenario for donor impact and institutional priorities.
- **Public Institutional Data (News, Rankings, Sentiment)**
  - Outward-facing information about the institution, gathered through the website, ranking and ratings, media coverage, public statements, social media, government data, and public reports. This is essentially how others see the organization and can be a driving force or hindrance for giving.

Five-Step Implementation Playbook

Used to make the Four Data Frontiers actionable, this sequence of steps will secure quick wins that demonstrate progress and encourage buy-in.

- **Map current data by quadrant.**
    - Determine the existing information stream and plot it into the four quadrants. This analysis will identify weak or blind spots.
- **Establish a master ID across systems.**
    - This ID will link all the records and data across the different systems. The easiest choice is an email address and not a unique constituent ID, since some of the external datasets are likely to match faster on a common email.
- **Launch trigger pilot.**
    - Choose a simple pilot, maybe one cross-quadrant, and create an automated data trigger for it.
    - Ex: Configure a rule in your donor information system where *if* an alum updates their job title on LinkedIn to CEO or another C-Suite title, that person is flagged as having a major gift proposal and is labeled a priority outreach. You can add more rules and make it more complex to suit your needs.
- **Automate data hygiene and monitoring.**
    - With a successful pilot, start thinking of how to scale and sustain the system. This could mean automating data cleanup. Establish clear governance practices to keep things on track, and set up regular data syncs.
- **Layer in AI insights with governance guardrails.**
    - Add an AI-driven tool to supercharge insight generation and move to action.

Advancement AI Ethics Maturity Framework

The AI Ethics Maturity Spectrum describes the policy and mindset needed for ethical AI governance in mission-driven institutions.

- **Level 1: Adhoc**
    - There are no rules on AI usage because it involves isolated, often untested tools like ChatGPT or Grammarly. Leaders either don't know or don't care, creating a lot of ethical risk.
- **Level II: Aware**
    - Ethics is relatively reactive, responding to a pointed need for policy. Leadership is interested in the use of AI but remains hands-off. The nonprofit begins to define what *shouldn't* happen, but not yet what should happen or how.
- **Level III: Defined**
    - Policies are created, with teams following approval workflows for tool adoption with clear oversight or an ethics committee.
- **Level IV: Policy Guided**
    - AI governance is organized, and an ethics review is often needed for project intake. Intentional Legitimate Interest Assessment (ILIA) is often adopted alongside board oversight.
- **Level V: Embedded Culture**
    - Ethics is built in and requires ongoing learning. The nonprofit has regular third-party audits, updates its policies every year, and builds ethical reflexes into hiring, training, and buying tools.

Five-Phase Design Thinking

This model recognizes that AI adoption is human-centered and iterative rather than a traditional linear rollout with a clear beginning and end.

- **Empathize: Ground AI in Human Voices**
  Include frontline fundraisers, donor relations professionals, donors, alumni, and other people who will benefit from, interact with, or be affected by the AI tool or system you are designing. The goal is to map pain points, expectations, and hopes before jumping to tech solutions.
- **Define: Articulate the Problem That Your AI System Will Solve**
  Synthesize the empathy findings to pinpoint the core needs and create a problem statement to solve them. Focus on the human need and not a technical issue.
- **Ideate: Generate Responsible AI Use Cases and Creative Solutions**
  Doing the ideation phase as the third step is essential to better understand how people feel and what the human problems are before you start solving them. The output of this phase is likely to be a list of solution concepts to advance to prototyping.
- **Prototype: Design Small-Scale Pilots with Checkpoints and Ethical Guardrails**
  For AI integration, this means creating a testable and tangible version of the solution with the mindset of building to think. Checkpoints and guardrails should be built into the prototypes to understand when to advance to the next step, what is being tested for, what data is being collected, and when

to pull the plug on the project. Another key theme here is to fail fast. It does not mean to move fast or haphazardly, but rather the ability to fail fast, learn from that, and move to the next iteration.

- **Test: Gather the Feedback, Measure Trust, and Refine Your Work**
  The prototype comes to life and engages with the reality of day-to-day usage. This means trying out the AI solution or pilot in the real world, but in a controlled environment with different scenarios to collect feedback from everyone involved. Success in advancement settings is measured in trust, adoption rate, and results.

Seven Skills for AI-Enabled Professionals (BHEF 2025)

- **AI Literacy:** A baseline understanding of key AI concepts and tools, specifically understanding how AI works, its limitations, and its capabilities, rather than seeing it as a "black box."
- **Data Literacy:** The ability to understand and work with data. Data is only beneficial if it can be analyzed and applied to make informed decisions.
- **Critical Thinking and Creativity:** An understanding of where machines fall short and applying human creativity to problem solve based on AI output.
- **Ethics and Responsible AI Use:** A solid foundation in data ethics, privacy, and an organization's particular AI governance and policies, including concepts like bias, consent, and transparency.
- **Digital and Computational Skills:** The ability to adapt to advancements and navigate the automated, digitized systems.

- **Collaboration and AI Era Communications:** Adjusting interpersonal skills to collaborate seamlessly with AI.
- **Adaptability and Continuous Learning:** Keeping up with the constant, transformational changes and adapting to the best tools.

Five Core Capabilities

- **Insight:** Develop the ability to generate data-driven foresight and insights. From an AI lens, this would include predictive modeling, natural language processing, or even enhanced reporting.
- **Activation:** Commit to deploying these insights in day-to-day work.
- **Governance:** Establish the processes and structures that oversee the use of AI to ensure alignment with institutional values and ethical principles.
- **Enablement:** Ensure that the infrastructure and guidelines can position the team to be AI-enabled, including training programs and support services.
- **Measurement:** Evaluate the metrics to determine impact and learn from what the data tells us. This involves clearly defining what success looks like and defining ways to monitor and evaluate it.

Instead of thinking about AI as a separate tech project, imagine it as strengthening these core muscles of the organization. Equally important, these capabilities cut across the traditional organization chart and help break silos and bring stakeholders together.

Three Structural Pathways for Scaling AI Talent and Experimentation.

Model 1: Hub and Spoke (Center of Excellence)

A central team (hub) concentrates AI expertise, setting the standards, providing guidance, developing core tools or models, and handling the most complex tasks (like building a custom AI model). The spokes are the various units or departments that the hub supports.

Model 2: Community of Practice

This networked approach relies on voluntary communities of practice rather than a strict hierarchy. The idea is to create forums where people interested in AI can share their knowledge, collaborate on ideas, and spread best practices across silos. These could be other units within your advancement team, others across your nonprofit or campus, or even others in the industry who have similar roles at other institutions.

Model 3: Innovation Lab

The third model consists of creating an AI lab, often with a corporate partner, and uses that as the dedicated virtual, physical, or organizational space for experimentation and prototyping, such as piloting new ideas in a low-risk environment. The key advantage is that an innovation lab creates a close environment to ideate, pilot, test, fail fast, and pivot, then scale when things work out.

12-Week Upskilling Sprint

- **Weeks 1 and 2: Foundations and Framing**
  Kick off with training on AI basics and your organization's AI vision. This could involve a series of interactive workshops covering ethical

considerations and case studies from other institutions. An executive sponsor should articulate why the institution is adopting AI, address fears, and build excitement around it.

- **Weeks 3, 4, and 5: Skills Pathways by Role**
  Using internal talent (if available), external trainers, or online platforms like LinkedIn Learning, assign staff to role-specific learning paths. The idea here is not to turn every team member into a computer scientist, but rather to give each person a concrete sense of how AI can augment their work.
- **Weeks 6, 7, and 8: The Hands-On Sprints**
  Shift gears here from learning to doing hands-on projects, preferably fundraising-related and cross-functional.
- **Weeks 9 and 10: Synthesis and Upskilling Sprint "Showcase"**
  Share results from the project sprints and consolidate lessons. Conduct a demo day, almost like a science fair, and bring outside audiences, like institutional leadership, representatives from other nonprofits, or corporate partners.
- **Weeks 11 and 12: Integration into Workplans, Job Descriptions, and Next Steps**
  The final phase ensures that AI is not a one-off burst and establishes how each participant can create a personal development plan or work plan to integrate the new skills or follow-up actions. On the institutional level, this is a good time to make a permanent support group, perhaps an AI working group, to help promote a culture of consistent upskilling within your organization.

CASE's Generative AI Findings as a Sector Adoption Baseline

CASE found that while most advancement professionals are personally eager to adopt AI, their teams and leaders often take a laissez-faire or hesitant approach. Senior leadership is usually not outright prohibiting AI exploration, but neither are they actively promoting innovation, resulting in stagnation by default.

Sector-Level AI Risk Taxonomy

When considering AI adoption across an entire sector, like the nonprofit or higher-ed advancement sector, it is helpful to categorize the risks at a macro level.

- **Reputational Risk:** The danger of trust and credibility loss due to AI misuse or failures. Public controversies could erode trust.
- **Operational Risk:** Risks of disruption to operations or service quality from AI.
- **Ethical Risk:** The potential for harm or injustice resulting from AI decisions or data usage.
- **Compliance/Legal Risk:** For example, HIPAA compliance. Or compliance with emerging AI regulations.
- **Strategic Risk:** The risk of mission failure or missed opportunities due to AI strategy missteps. If an institution falls behind in AI capability, it may lose relevance.

Toolkit: Readiness Signals for Chief Advancement Officers, Presidents, and Boards

The following is a toolkit of signals and questions for organizational leaders to assess where they stand and where to focus. You can use these prompts as a checklist in leadership meetings or strategy sessions:

- **Board Engagement**
  - **Signal:** The board has explicitly discussed AI opportunities and risks within the last year.
  - **Questions:** If AI hasn't made it onto the board's agenda, it's a sign that the institution isn't treating AI as a strategic issue. Boards should inquire: Do we have an AI strategy or policy in place? What is our exposure to AI right now? A proactive board might even form a board subcommittee on technology or require periodic AI risk reports.
- **Executive Ownership**
  - **Signal:** A specific senior leader or cross-functional committee is charged with overseeing AI initiatives and governance.
  - **Questions:** Can the president or chief advancement officer name who is responsible for the AI strategy? If it's diffused or "no one," that implies low readiness. Look for signs like the appointment of an "AI lead" or the establishment of an AI task force.
- **Policy Framework**
  - **Signal:** The organization has at least a preliminary AI policy or set of guidelines or guiding principles in place, even if this is done informally.
  - **Questions:** Is there a written document or an agreed-upon set of guidelines about how staff should (or shouldn't) use AI tools?

- **Staff Activity (Visibility)**
  - **Signal:** Staff use of AI is known to leaders.
  - **Questions:** Can you list some AI tools or use cases people on your team have tried?
- **Training and Culture**
  - **Signal:** There is an ongoing effort to train staff on AI-related skills.
  - **Questions:** Did the nonprofit send its staff to AI conferences or plan its own training? Or provide resources for learning (like CASE or AFP webinars or online courses)? Are the team meetings occasionally featuring demos of new tech? Is there a climate of fear and skepticism?
- **Ethics and Risk Protocols**
  - **Signal:** Before launching any AI-related project, there is a process, whether formal or informal, to review risks involving multiple stakeholders (e.g., IT, advancement, legal).
  - **Questions:** What steps do you take to evaluate an AI tool before using it?
- **Technology Infrastructure**
  - **Signal:** The IT team has integrated or has planned for AI capabilities in the existing systems (CRM, databases, etc.) and has measures for data security in those contexts.
  - **Questions:** Does your data environment support AI, and is your CRM offering AI capabilities?
- **Early Wins and Learnings:**
  - **Signal:** Your organization can point to at least one or two pilots that have been completed and documented, whether successfully or not.

    - **Questions:** What have we learned from our AI experiments so far?
- **Stakeholder Sentiment:**
    - **Signal:** Donors and alumni have shared their thoughts about the use of AI.
    - **Questions:** Have you surveyed your board or donors on the use of AI? Did you have a conversation with your community in a town hall?

Exhibit A:
"Comprehensive Donor Intelligence Report Framework"

Instead of doing prompt engineering and designing it ourselves, we asked ChatGPT to take the role of a prompt designer and help us design a simple prompt that can produce publicly available data on a prospective donor. We then kept refining the request by sending the prompt back to ChatGPT and iterating on it multiple times, adding perspective and examples. We ended up asking for a detailed "Super Prompt" that was accompanied by the Association of Fundraising Professionals' Code of Ethics, Donor Bill of Rights, and four additional organizations' codes of ethical standards to help make the prompt more in line with an advancement office.[1, 2] As technology advances, the prompt itself might be a bit dated, but the approach by which we produced it can be easily replicated. This is meant as an illustrative example only, a reusable template

---

1 "Code of Ethical Standards," Association of Fundraising Professionals, updated December 2023, https://afpglobal.org/ethicsmain/code-ethical-standards.

2 "Donor Bill of Rights," Association of Fundraising Professionals, https://afpglobal.org/donor-bill-rights.

for nonprofit cultivation, stewardship, and campaign alignment.

> You are a donor intelligence analyst. Using deep research and only public sources, produce a comprehensive, decision-ready profile on [DONOR NAME, COMPANY NAME, CITY, STATE]. The report must support cultivation, stewardship, and campaign alignment for [INSTITUTION NAME] [PRIORITY 1, PRIORITY 2, PRIORITY 3], identifying value alignment, partnership opportunities, risks, and practical next actions.

Global Research Rules

- **Recency and Relevance:** Prioritize findings from the last five years. Include earlier items only if they clarify philanthropic evolution, leadership arc, or institutional alignment.
- **Identity Resolution:** Confirm identity using at least three indicators (education, employer, geography, board roles). Flag ambiguity with a confidence score.
- **Source Types (Public Only):** Bloomberg, Forbes, Crunchbase or similar organizations; Guidestar/Candid, IRS 990s; SEC/10-K filings; major media; institutional releases; LinkedIn; Giving Pledge; conference bios; award citations.
- **Cross-Verification:** Every financial, philanthropic, or governance detail is verified by at least two independent sources
- **Privacy and Compliance:** Include spouse/family information only if publicly acknowledged. Report city-level location only; exclude private data.

- **Evidence and Citations:** Every nonobvious claim must carry a citation (e.g., Source—Publisher, Year, URL).
- **Clarity and Integrity:** No speculation. Use "Not publicly available" or "Conflicting reports (Low/Med/High confidence)" when applicable.
- Bias and Fairness Safeguards
- **No Demographic Inference:** Do not infer values, motives, wealth, or ideology from race, ethnicity, religion, gender, nationality, school prestige, or neighborhood.
- **Facts vs. Interpretation:** Clearly separate cited facts from interpretations and recommendations; tie recommendations directly to evidence.
- **Source Weighting:** Prioritize primary sources (IRS 990s, SEC filings, foundation reports, direct quotes). Corroborate media claims; label unverified items.
- **Materiality Test:** Include controversies only if reputationally relevant; describe neutrally and cite responses when available.
- **Political Neutrality:** Report political giving descriptively only; no character judgments.
- **Scoring Discipline:** Any alignment score ≥4 or ≤2 requires at least two cited indicators; otherwise, score "3—Mixed/Unknown."
- **Transparency:** Include "Unknowns/Not Publicly Available" and a confidence rating in each section.

Output Deliverables

A. Executive Brief (≤250 words)

   a. Who they are; top cause areas; alignment with institutional pillars; largest opportunity; largest risk; "next three moves" (actionable steps).

B. Comprehensive Report (Sections 1–10)

C. Appendices

   a. Full citation list (alphabetized) and CSV-ready tables of major gifts, boards, and networks.

Comprehensive Report Outline

- **Section 1:** Background and biography; legal/preferred name; titles; aliases; education and degrees; career timeline and milestones; current roles and boards (corporate, nonprofit, advisory); estimated age; residence (city/state/country); publicly acknowledged family members with philanthropic activity.
- **Section 2:** Philanthropic interests and motivations; primary cause areas; 3–5 direct quotes reflecting values or worldview; giving philosophy (naming, visibility, outcomes); vehicles: foundations/DAFs; giving archetype (e.g., tech-enabled, equity-driven).
- **Section 3:** Giving history and financial indicators (year, donor vehicle, recipient, gift type, purpose, approx. amount, source[s]); secondary support (fellowships, in-kind, prizes); estimated net worth (Forbes/Bloomberg, with date); foundation assets, public holdings; rationale for naming or impact design (if stated).
- **Section 4:** Specific connections to the institution; past collaborations or affiliations; quotes referencing the institution's faculty, programs, or labs; grants, in-kind donations, honorary degrees, and visits; alumni or partner relationships.
- **Section 5:** Alignment with institutional mission; heatmap of alignment (1–5 scale) for each priority or pillar; 3–6 opportunity concepts (e.g., endowed

chair, applied lab, joint research); messaging snippets using the donor's own phrasing.

- **Section 6:** Broader institutional and peer engagement; comparable investments at peer institutions; advisory or leadership roles at similar organizations; overlapping networks or shared affiliations (table if ≥5 overlaps).
- **Section 7:** Strategic engagement recommendations; 3–5 cultivation strategies (lectures, roundtables, site visits, showcases); 2–3 naming frameworks prioritizing mission over vanity; 2–3 gift framings (e.g., seed + match; endowed + applied); draft outreach email (120–150 words) from the most credible sender.
- **Section 8:** Media, speeches, and thought leadership; keynotes, interviews, and op-eds; media tone and visibility themes; communication style analysis.
- **Section 9:** Political and policy engagement (optional); summary of bipartisan contributions (> $10K/cycle); PAC/issue advocacy (if public); policy board participation.
- **Section 10:** Risk factors and sensitivities; preferences for anonymity or quiet giving; reputational or ethical sensitivities (cited neutrally); leadership style, decision cadence, and expectation management; media or peer response to past gifts.

Table Specifications (CSV-Ready)

- Major Gifts Table: See Section 3.
- Boards and Affiliations Table: Org | Role | Start–End | Domain | Source(s).
- Political Giving Table (if used): Year | Recipient/Committee | Amount | Party/Issue | Source(s).

- Network Overlaps Table: Person/Org | Overlap Type | Relevance | Source(s). Scoring and Confidence. Each section concludes with a Confidence Rating: High/Medium/Low.

Note: Include an "Unknowns/Not Publicly Available" list for transparency.

APPENDIX

Here is a catch-all addition to your gift acceptance and advancement policy on the use of AI in advancement.

USE OF ARTIFICIAL INTELLIGENCE IN UNIVERSITY ADVANCEMENT

Recognizing the importance of the rapid advancement in technology, and in order to streamline and enhance the efficacy of our advancement team, this policy authorizes the use of artificial intelligence (AI) tools within the university advancement operations. These tools are designed to aid in streamlining processes and potentially in data collection, analysis, and decision-making related to advancement activities. The utilization of AI will comply with all applicable laws, regulations, and university policies, including privacy and data protection, to maintain the trust and respect of our donors, alumni, and community. University advancement is encouraged to identify, evaluate, and implement AI applications that can support our mission of maintaining and growing the resources available for the university's educational, research, and public service activities. All AI-related technology implementations will be reviewed by the president (or a designate), provost (or a designate), general counsel, chief information services officer, and vice president for university advancement to ensure alignment with our core values and ethical guidelines. The advancement committee of the board shall also periodically provide oversight and review the effectiveness and ethical implications of these AI tools.

ABOUT THE AUTHORS

Joe Manok is the vice president for advancement at Clark University and the founder of GlobalPhilanthropy.ai. A certified fundraising executive (CFRE), Joe has also earned an executive certification in strategy and innovation at MIT's Sloan School of Management.

Since 2014, Joe has held senior-level positions in the Office of Resource Development at MIT, serving as senior director of philanthropic partnerships and as director of global initiatives. At MIT, Joe was part of the principal giving team and successfully closed multiple principal and transformational gifts as part of MIT's $6.28 billion *Campaign for a Better World.*

Before joining MIT, Joe was a rising-star leader within the Office of Development at the American University of Beirut (AUB), serving most recently as associate director of development for major gifts and advancement services in New York. He had broad responsibilities at AUB, including managing a portfolio in North America; directing major, principal, and planned giving; and overseeing many elements of alumni programming. He personally re-engaged hundreds of disengaged alumni and helped AUB launch its first online alumni community. In 2007, AUB honored Joe with the President's Service Excellence Award, and in 2011, he was recognized by the Council for Advancement and Support of Education (CASE) with the Rising Star Award for his consistently high level of professional achievement and commitment to the advancement profession.

A computer scientist by trade, Joe is a leading expert on integrating artificial intelligence into philanthropy to modernize higher education operations. He is a frequent speaker at national higher education conferences and has published extensively referenced research on trends in principal giving, global philanthropy, and transformational gift conversions.

David Woodruff is a consulting partner at Marts & Lundy with over 40 years of fundraising expertise. Having joined the firm in 2022 following his role as associate vice president and COO of resource development at MIT, he led development operations and oversaw a successful $6 billion *Campaign for a Better World*. Under his leadership, the team surpassed $500 million in philanthropic commitments for eight consecutive years and has been positioned for sustained growth post-campaign.

As the executive director and COO of development at Massachusetts General Hospital, David oversaw the $1.5 billion *Campaign for the Third Century of MGH Medicine*. He also served as dean for resource development and chief development officer at the Harvard T. H. Chan School of Public Health, where he expanded annual fundraising results from all sources and advanced the school's global mission and impact.

A certified fundraising executive, David specializes in strategic planning, organizational development, and securing principal gifts to help nonprofits achieve transformative growth. He is a passionate champion for nonprofits and committed to fostering organizational development, managing and mentoring talent, planning and executing strategy, and building sustainable business operations. In 2025, David received the Lifetime Achievement in Fundraising Award from the Massachusetts chapter of the Association of Fundraising Professionals. He recently launched Holdfast Coaching, LLC, an executive coaching practice assisting nonprofit leaders in their professional growth and development.

www.ingramcontent.com/pod-product-compliance
Lightning Source LLC
LaVergne TN
LVHW022159200926
848481LV00006B/1789